Broadband Telecommunications and Regional Development

Broadband is one of the most transformative technologies of the twenty-first century, yet our understanding of its regional impacts remains somewhat rudimentary. Not only are issues of broadband pricing and speed relevant in this context, but the overall quality of service for broadband can often dictate its impacts on regional development. This book illuminates the regional impacts of this pervasive and important technology.

The principal aim of this book is to deepen our understanding of broadband and its connections to regional development. First, it uses a geospatial lens to explore how the relationship between broadband and regional development influences access to technology platforms, dictates provision patterns and facilitates the shrinkage of space and time in non-uniform and sometimes unexpected ways. Second, it provides a comprehensive guide that details the strengths and weaknesses of publically available broadband data and their associated uncertainties, allowing regional development professionals and researchers to make more informed decisions regarding data use, analytical models and policy recommendations. Finally, this book is the first to detail the growing importance of broadband to digital innovation and entrepreneurship in regions.

This book will be of interest to regional development professionals and researchers in economics, public policy, geography, regional science and planning.

Tony H. Grubesic is a professor in the College of Public Service & Community Solutions at Arizona State University, USA.

Elizabeth A. Mack is an assistant professor in the Department of Geography at Michigan State University, USA.

Routledge Advances in Regional Economics, Science and Policy

Broadband Telecommunications and Regional Development

Tony H. Grubesic and
Elizabeth A. Mack

LONDON AND NEW YORK

First published 2016
by Routledge

4 Park Square, Milton Park, Abingdon, Oxon OX14 4RN
605 Third Avenue, New York, NY 10017

First issued in paperback 2017

Routledge is an imprint of the Taylor & Francis Group, an informa business

British Library Cataloguing in Publication Data
A catalogue record for this book is available from the British Library

Library of Congress Cataloging in Publication Data
Grubesic, Tony H.
Broadband telecommunications and regional development / Tony H.
Grubesic and Elizabeth A. Mack.
Includes bibliographical references and index.
1. Telecommunication policy—United States. 2. Broadband communication
systems—United States. 3. Internet service providers—United States.
4. Regional planning —United States. I. Mack, Elizabeth A. II. Title.
HE7781.G78 2015
384.30973—dc23
2015013029

ISBN 13: 978-0-8153-4725-5 (pbk)
ISBN 13: 978-1-138-01391-9 (hbk)

Typeset in Times New Roman
by Swales & Willis Ltd, Exeter, Devon, UK

Contents

Figures

Tables

Preface

Our motivation for writing this book is a simple one. We both like broadband. We like to use it, think about it, talk about it and write about it. Ironically, it is the "writing about it" part that we have found most difficult. There are several reasons for this. First, broadband is so popular and important that almost every discipline has something to say about it, including engineering, planning, economics, political science, policy, planning, history and commerce. This means that tracking down literature and synthesizing the current state of broadband affairs is inherently difficult and time consuming. Second, almost everybody has an opinion on broadband and, for whatever reason, nobody is shy about sharing these opinions. This includes general musings on providers, broadband costs, user experience and how broadband could be better. In other words, everybody is an expert.

Our goal in writing this book was fairly modest: to identify how geography acts as the connective tissue between many of the technical, social, economic and planning issues associated with broadband, and to highlight the potential role of broadband in regional development. Readers should keep this goal in mind because we make no attempt to cover the entire field of broadband, it is simply too diverse and too complex. That said, we do explore several facets of broadband that do not receive enough attention, including data uncertainty, broadband for business and entrepreneurship. We also provide an overview on policy, the spatial distribution of broadband provision and a basic summary of broadband technologies (written for non-engineers).

This book should appeal to a wide range of scholars and professionals working in the fields of geography, public policy, community development and planning. We also believe that this book would be ideal for instructors tasked with developing and teaching a course on telecommunication geography, the digital divide, or as supplementary material in a class focused on broadband telecommunications policy and regional development for the United States.

New technologies, applications and changes to broadband policy emerge quite regularly. Our work is ongoing. Regardless, we hope that readers enjoy this book as much as we liked writing it.

Tony H. Grubesic and Elizabeth A. Mack
Philadelphia, Pennsylvania
March 2015

Acknowledgements

In the course of writing this book, we have profited from our contact with many colleagues, scholars, researchers and students. So many in fact, that we cannot hope to acknowledge them all. Nevertheless, we would like to express our thanks to Ran Wei (University of Utah), Jake Nelson (Arizona State University), Fangwu Wei (Drexel University), Luc Anselin and Sergio Rey (Arizona State University; GeoDa Center for Geospatial Analysis and Computation), and Yasuyuki Motoyama (Ewing Marion Kauffman Foundation).

1 Introduction

Transformative broadband

> Broadband access is the great equalizer, leveling the playing field so that every willing and able person, no matter their station in life, has access to the information and tools necessary to achieve the American Dream. (Michael Powell, 2010, 1).

This quote from former Federal Communications Commission Chairman (FCC) Michael Powell and current president and CEO of the National Cable & Telecommunications highlights the enduring importance of high-speed or broadband internet connections to people, businesses and regional competitiveness. Unfortunately, over twenty years after the Internet first became available to the public in 1995 (Abbate, 2000), little is actually known about how people and especially businesses use broadband to become more competitive. In fact, very little is known about the link between businesses, broadband and their combined impact on regional development. These gaps in our knowledge base pose several questions. What do we really know about broadband? What is important to know about broadband? How does broadband impact business and regional competitiveness? How can research deepen our understanding about broadband and its social, economic, geographic, cultural and political effects?

These questions are the motivation and starting point for this book. Although broadband is not necessarily a new technology, the innumerable innovations and economic impacts associated with the Internet make it one of only twenty-four general-purpose technologies (GPTs) developed since 9000 bc (Lipsey Carlaw, & Bekar, 2005). This places the Internet in a unique class of inventions that have forever changed the development trajectory of humankind. Further, while the ubiquity of internet content (music, video clips, etc.) lead many to believe that the Internet is available and consumed everywhere, this assumption is erroneous. The Internet is *not* ubiquitous and tremendous disparities in availability persist globally, even within the United States.

With this backdrop in mind, the purpose of this introductory chapter is three-fold. Our primary goal is to provide readers with a detailed overview of the research questions to be addressed in this book and delineate pathways for their exploration. The second purpose of this chapter is to provide readers with a foundational vocabulary pertaining to broadband. We anticipate that those interested in this topic will have a diversity of backgrounds and experience, ranging from public policy, economics and sociology, to geography, regional science, communications and planning, amongst others. Thus, this chapter provides a primer on key terms, such as access, accessibility, availability and affordability, and adoption. Finally, although most readers will be aware that the public discourse on broadband consists of important economic, political, cultural, planning, engineering, demographic and socio-economic issues, many are unlikely to be aware of how geography acts as the connective tissue between all of these domains. Thus, this chapter sets the stage for pursuing answers to important broadband-related questions through a spatial lens, highlighting the importance of geospatial context and its relevance to a field (i.e., broadband) traditionally dominated by economists and public policy scholars.

Inequities

Prior research on broadband availability has found that specific populations are consistently at the fringes or at risk of falling into the digital divide. These studies highlight that low-income households (Martin & Robinson, 2007), people of African–American, Hispanic and/or Native American descent (Prieger & Hu, 2008), those living in rural areas (Strover, 2001), the disabled (FCC, 2010) and older populations (Eastman & Iyer, 2005) are frequently without high-speed broadband connections. More recent work on the spatial distribution of broadband highlights some important nuances to this divide, including the presence of broadband coverage gaps in major metropolitan areas (Grubesic & Murray, 2002; Grubesic, 2006). These local, spatial facets are important to acknowledge because aggregate, national-level studies of the digital divide are often too coarse to reveal spatial disparities in broadband access within cities, suburban and exurban areas (Grubesic & Murray, 2004).

One of the more interesting and challenging aspects of broadband is that its definition is a moving target. For the last several years, the FCC (2010) defined broadband as download (i.e., to the customer) speeds of at least 4 megabytes per second (Mbps) and upload (i.e., to the customer) speeds of at least 1 Mbps (4/1 Mbps; FCC, 2010). However, beginning in January 2015, these speeds were increased to 25 Mbps/3 Mbps (25/3 Mbps).[1] Based on this threshold, the FCC (2015) estimates that 55 million Americans lack broadband, but only 8% of urban residents lack access to the 25/3 Mbps service. The FCC (2015) also suggests that rural portions of the U.S. are underserved at all speeds, with a whopping 20% of rural residents lacking access to 4/1 Mbps service. These gaps between rural and urban areas have remained fairly persistent in the U.S. over the past decade, regardless of how broadband was defined.

Given the growing reliance of people and businesses on internet-enabled technologies and the perceived inequities in broadband infrastructure between urban, suburban, exurban, rural and remote areas, the private sector and government continue to expend an enormous amount of time, energy and money on rolling out telecommunications infrastructure to consumers. These efforts include national policy initiatives to promote competition, such as the Telecommunications Act of 1996 (*96 Act*); recent efforts for promoting equity, affordability and access in the National Broadband Plan (FCC, 2010); the need for better and more detailed broadband data (Grubesic, 2012); more research on the impacts that broadband has on business competitiveness; and what these business impacts mean for regional and national competitiveness.

One of the most pressing issues facing broadband is that telecommunications markets are becoming less competitive and more monopolistic, obliging Americans to pay higher rates for lower broadband speeds when compared to other consumers around the world (Crawford, 2013). This picture is not likely to improve anytime soon, particularly given the continued consolidation of large telecommunications conglomerates. The most recent mega-deal (~$45 billion) between providers is currently underway, where Comcast, the nation's largest cable provider, is attempting to purchase Time Warner, the nation's second largest cable provider. Figure 1.1 highlights the geographic footprint of the service areas for both providers. In a recent analysis conducted by GeoResults (Ulanoff, 2014), it was estimated that the Comcast service area covers nearly 11.6 million businesses and 46 million residential households, while Time Warner covers 6.2 million businesses and 25.7 million residential households. If this deal is approved,

Figure 1.1 Proposed Time Warner + Comcast merger.
For a full colour version of this image, please see www.routledge.com/books/details/9781138013919/

Source: Georesults

the new mega-company would maintain a cable monopoly over one-third of the United States and 70 million households. In effect, this would make Comcast the gatekeeper for six of every ten cable boxes in the U.S. (Clark Estes, 2015), potentially allowing them to control the distribution of streaming content, dictate pricing, etc. Their track record in this area is already questionable, recently forcing Netflix to pay for peering to ensure a more fluid delivery of content to Comcast subscribers (Wyatt & Cohen, 2014).

Another conundrum for U.S. broadband policy related to the Comcast/Time Warner merger is net neutrality. This term refers to the idea that all internet flows be treated equally by internet service providers (ISPs) and governments; free from discrimination by user, content, site, platform, application, equipment or mode of communication (Wu, 2003). More importantly, there is no discrimination in prices charged. This is important because it means all people and businesses have equal access to internet content, provided that the infrastructure is available to them and they are able to use it. Ironically, as the FCC prepared to pass a net neutrality plan in February 2015, Michael Powell (quoted above), suggested that the proposed net neutrality rules are a "sad example of unreasoned decision-making," and that "watching the president of the United States direct the FCC to adopt a very specific regulatory result, I think . . . was shocking and put the commission in an untenable position" (Weil, 2015, 1). This is an odd statement, coming from somebody who just five years earlier touted broadband as a vehicle to achieve the American Dream.

While net neutrality is not the primary focus of this book, this issue is important to acknowledge because it influences use of broadband and the Internet. Without net neutrality, ISPs can discriminate between users, content, sites, platforms, applications and modes of communication. Sadly, on this uneven playing field, broadband quality of service, user experience and available content is influenced by the size of one's checkbook. This is very different from an egalitarian broadband ecosystem that the U.S. needs (and deserves) to maintain a competitive edge in the global information economy.

A primer

Before proceeding to a description of the primary foci of this book, it is necessary to highlight some subtle differences in the vocabulary used to discuss broadband. This is important because the lack of rigorously structured vocabulary for referencing the many facets of broadband services and their consumption can prevent meaningful dialog between economists, planners, sociologists and other stakeholders in the policy, development and analytics communities. For example, there are major differences between broadband *access*, *accessibility*, *availability*, *affordability* and *adoption*. It is also critical to highlight the importance of a spatial perspective on broadband issues given the significant geographic nuances in infrastructure capabilities and infrastructure distribution.

To start, a core focus of the digital divide literature concerns disparities in broadband access (Compaine, 2001; Norris, 2001). *Access* generally refers to the ability of a household, business or individual to receive broadband services. Access is typically captured in terms of proximity or convenience. For the purposes of this book, our primary focus concerns the spatial dimensions of access and identifying how geography plays a critical role in either facilitating or complicating broadband access. For example, it is well known that xDSL access is strongly associated with household distance from a central office or a remote digital subscriber access multiplexer (Grubesic, 2008a). As detailed by Grubesic (2008a), xDSL services are generally unavailable to households or businesses that are more than 3.6 km (~18,000 ft.) from a central office. Simply put, this is a measure of access.

Conversely, *accessibility* refers to the portions of the Internet that can be viewed or interacted with once a user is on the network. High levels of accessibility are generally found in places (and on networks) where internet content is not censored (e.g., the United States). Low levels of internet accessibility are typified by places such as China. For example, the Chinese government prohibits the consumption of some content (e.g. material that subverts state power, pornography and obscenity) via the "Golden Shield" program (Liang & Lu, 2010). Of course, this is not to say that all content on the Internet is 100% accessible for users in locations where censorship is non-existent, some material requires paid subscriptions, passwords or other types of affiliation-based permissions.

Availability relates to the extent that broadband services can be received within a reasonable waiting time. In the U.S., a reasonable waiting time is loosely defined as 7–10 days (Grubesic, 2012), which corresponds to the typical wait time for an installation appointment from a broadband provider with facilities in the area (NTIA, 2011). *Affordability* refers to the extent to which a household, business or individual has the resources to pay for broadband services. To be clear, affordability is significantly different from *pricing*, which refers to the actual cost of broadband service. For example, a basic broadband package priced at $60 per month is probably more affordable to a household with a yearly income of $100,000 than it is to a household with an income of $10,000 per year. Finally, from a theoretical perspective, *adoption* is generally focused on factors that influence decision-making processes for households, businesses or individuals. For example, where broadband (and technology more generally) is concerned, households with higher income and education levels tend to be early adopters (Rogers, 2003; LaRose, Gregg, Strover, Straubhaar, & Carpenter, 2007; Whitacre, 2008), but racial and ethnic characteristics and urban/rural status also play a role (Horrigan, Stolp, & Wilson, 2006; Whitacre & Mills, 2007).

There is also some confusion surrounding the differences between broadband provision, demand and diffusion. Mack and Grubesic (2009), note that broadband *provision* is the physical act of making some type of broadband infrastructure available to potential subscribers within a region. The *demand* for broadband, however, is contingent upon a complex web of factors, including price, service quality, consumer attributes (e.g. demographic profile, income levels and location) and technology alternatives (e.g. narrowband). For example, affluent suburban communities often display a greater demand for broadband services than disenfranchised and impoverished inner-city locations (Grubesic, 2004).[2] The diffusion of broadband infrastructure and associated services is more complex and relates to the interaction of four factors (Rogers, 2003): 1) the innovation, 2) communication channels, 3) time and 4) a social system. As detailed by Whitacre (2008), if the innovation is broadband access, then the communication channels can consist of conversations with work colleagues and/or neighbors or local advertising campaigns. The temporal component of broadband, at least where provision is concerned, is rooted in supply-side considerations such as return on investment for the providers, local population density, business mix, income, race, age and geography (Grubesic, 2003; Prieger, 2003; Strover, 2003; Grubesic, 2006; Grubesic, 2008b; Mack and Grubesic, 2009; Mack, 2014; Mack and Grubesic, 2014). Finally, the social system can be defined as all households or businesses within an area, since the adoption decision is made by each individual entity.

Aside from these basic vocabulary issues, many broadband research and policy efforts suffer from a superficial understanding of broadband data and their visual representations. This can, and often does, prevent the development of a meaningful telecommunications policy. Although maps and associated visual analytics are both powerful and convenient, maps can also lie (Monmonier, 2014). Figure 1.2, for instance, illustrates variations in geographic coverage for AT&T wireless services along the Central Oregon Coast, near the community of Waldport. For residents of Waldport, a quick glance at the information visualized in Figure 1.2 may (or may not) influence their decision to adopt AT&T as their wireless provider. More importantly, if analysts are conducting empirical work to identify

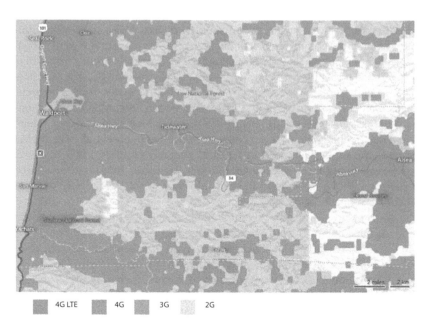

Figure 1.2 AT&T wireless coverage: Central Oregon coast, 2015.

Source: http://www.att.com/maps/wireless-coverage.html

gaps in wireless broadband provision, could the data and associated visualization presented in Figure 1.2 be used as a foundation for developing wireless policy at the local, state or national levels? Is the gradient of wireless data coverage as steep as it looks when one moves east from Waldport and the Pacific Coast? How and when were these data collected? Are these advertised speeds or realized speeds? How were the coverages interpolated?

These are all questions that merit attention and are relevant for both wireless and wireline data. Thus, as mentioned previously, the mechanics associated with collecting, tabulating, summarizing, synthesizing, representing, analyzing and visualizing broadband data and their spatial facets are exceedingly complex, full of abstractions and uncertainties. So too are the local administrative units used for representing broadband data – all of which have complex geographical manifestations. As a result, without understanding the geography of broadband, one cannot possible hope to understand broadband. This book has been written to help fill this void.

The path ahead

The principal aim of this book is to deepen our understanding of broadband and its connections to regional development through a spatial lens. This includes detailed discussions regarding infrastructure, data and policy issues from a geographic perspective. Another major focus of this book is broadband and businesses. This is important because current discussions of broadband and associated policies are overwhelmingly focused on broadband for personal use. This ignores issues relevant to business adoption and use of broadband, which have wide-reaching economic implications. Thus, if we are to understand the economic impacts associated with broadband, the business issue must be addressed more thoroughly. This includes entrepreneurial opportunities provided by the Internet, and the use of broadband by entrepreneurs to improve the visibility and competitiveness of new ventures.

While the content of this book discusses themes that impact regional economies globally, its empirical focus is the United States. This emphasis is important because although the United States was one of the first countries to deploy broadband, it has fallen behind. The U.S currently ranks 16th in the world for fixed broadband subscriptions (per-capita; OECD, 2014). Further, as detailed earlier, there are 55 million people in the U.S. who do not have access to 25/3 Mbps service and 20% of rural residents cannot access basic 4/1 Mbps broadband service (FCC, 2015). In short, the U.S. provides an amazingly diverse and dynamic study area with issues that are particularly salient for geographically expansive countries grappling with provision and policy issues (e.g., Canada, Brazil, Australia and India).

As highlighted above, the remaining seven chapters of this book provide an overview of broadband issues in a U.S context, highlighting the value of and need for data and research that considers the inherently spatial nature of broadband infrastructure and its link to regional development. Chapter 2 provides an overview on past and present broadband-related telecommunications policy.

This overview includes a discussion of the Communications Act of 1934; the breakup of AT&T and the creation of the Baby Bells; and the Telecommunications Act of 1996. A discussion of municipal broadband regulations and initiatives closes the chapter. This is important to address given recent efforts by legislators at the national level to overturn municipal restrictions and prohibitions of community broadband efforts.

An important feature of this book is using a geospatial lens for exploring issues of broadband technology, data quality, provision, access and equity. Chapter 3 provides a detailed description of current broadband infrastructure technologies, including digital subscriber lines (xDSL), hybrid fiber-coaxial cable (HFC), fiber to the curb/home/building/node (FTTx), wireless (e.g., Wi-Fi, WiMAX, Cellular) and satellite. The chapter includes discussion of a number of key limitations, spatial or otherwise, for each of these platforms.

Given the plethora of broadband data that is now available to the public and private sectors, the use and abuse of these data has soared in recent years. Chapter 4 provides an objective look at publically available broadband databases with a focus on the FCC Form 477 and National Broadband Map (NBM) data. The chapter introduces readers to important elements of spatiotemporal error and uncertainty inherent to these data and how these errors can impact analytical assessments. Strategies for mitigating these uncertainties are also discussed, allowing regional development professionals and researchers in economics, public policy, geography, regional science and planning to make more informed decisions regarding data use, analytical models and policy recommendations.

Chapter 5 provides comprehensive exploration of the spatial dynamics of broadband provision in the United States. Details on which metropolitan areas in the United States have the highest levels of broadband availability are highlighted, as well as significant interregional differences between metropolitan, micropolitan, urban and rural areas.

Chapter 6 begins a discussion of broadband and businesses using a conceptual framework of this relationship called the Broadband-Business Nexus (BBN). The BBN highlights the supply and demand sides of this relationship. A spatial analysis of the spatial relationship between broadband and business is then conducted using ZIP code area data.

An important sub-dimension of the BBN is digital innovation and its impact on new firm formation, which is one way of characterizing entrepreneurial activity. Chapter 7 is an extended discussion of the innovation opportunities provided by broadband and the Internet. This discussion of innovation includes a brief historical overview of device and application advances in the last several decades.[3] After the innovation overview, a spatiotemporal analysis of broadband and new business activity is conducted across and within six metropolitan areas using high-resolution data on new business formation and broadband provision. The analysis highlights the dynamic and highly contextual association between broadband and entrepreneurial activity within metropolitan areas. Again, these results underscore both the importance and growing need for more in-depth spatial analyses at the local level, rather than aggregate, national-level analyses.

Finally, Chapter 8 ties the related topics of the preceding chapters together and outlines a number of important, emerging challenges for broadband provision, policy and associated implications for advancing regional competitiveness in the U.S. and beyond. This includes issues related to infrastructure deployment, data reporting and availability, broadband use, broadband policy and broadband in the developing world.

In sum, the contents of this book highlight a small, but important subset of empirical results that help address questions on the impacts of broadband and regional development. As stated previously, we believe that a spatial perspective on broadband issues is absolutely critical to making informed policy decisions, mitigating gaps in broadband deployment, initiating digital literacy campaigns for people and small businesses and reversing the competition gap that is growing between places as broadband technologies and markets continue to evolve. Although *Broadband Telecommunications and Regional Development* does not have all the answers, we hope that readers will appreciate the attention drawn to the spatiality of broadband and that this book will encourage the private sector, research and policy communities to engage in pursuing solutions to the persistent gaps in broadband access, availability, affordability, access, use and competition that exist within the United States and so many other countries around the world.

Notes

1 The FCC uses speed tiers as well, The FCC also breaks broadband speeds into eight unique tiers: 1st Generation ($\geq$ 200 kbps), Tier 1 or greater ($\geq$ 768 kbps), Tier 2 or greater ($\geq$ 1.5 Mbps), Tier 3 or greater ($\geq$ 3 Mbps), Tier 4 or greater ($\geq$ 6 Mbps), Tier 5 or greater ($\geq$ 10 Mbps), Tier 6 or greater $\geq$ 25 Mbps), Tier 7 or greater ($\geq$ 100 Mbps).
2 For more details on broadband demand and its determinants, see Flamm and Chaudhuri (2007).
3 Core network advances are not discussed in depth for Chapter 7 because this information is addressed in Chapter 3.

References

Abbate, J. (2000). *Inventing the internet*. MIT Press.

Clark Estes, A. (2015). The Comcast-Time Warner cable merger may not happen. *Gizmodo*. Retrieved from http://tinyurl.com/q9sxju6

Compaine, B.M. (Ed.). (2001). *The digital divide: Facing a crisis or creating a myth?* MIT Press.

Crawford, S.P. (2013). *Captive audience: The telecom industry and monopoly power in the new gilded age*. Yale University Press.

Eastman, J.K., & Iyer, R. (2005). The impact of cognitive age on Internet use of the elderly: An introduction to the public policy implications. *International Journal of Consumer Studies*, *29*(2), 125–136.

Federal Communications Commission [FCC]. (2010). *Connecting America: The National Broadband Plan*. Retrieved from http://www.fcc.gov/national-broadband-plan

Federal Communications Commission [FCC]. (2015). *Broadband availability in America*. Retrieved from http://tinyurl.com/qxscpvs

Flamm, K., & Chaudhuri, A. (2007). An analysis of the determinants of broadband access. *Telecommunications Policy*, *31*(6), 312–326.

Grubesic, T.H. (2003). Inequities in the broadband revolution. *The Annals of Regional Science*, *37*(2), 263–289.

Grubesic, T.H. (2004). The geodemographic correlates of broadband access and availability in the United States. *Telematics and Informatics*, *21*(4), 335–358.

Grubesic, T.H. (2006). A spatial taxonomy of broadband regions in the United States. *Information Economics and Policy*, *18*(4), 423–448.

Grubesic, T.H. (2008a). Spatial data constraints: Implications for measuring broadband. *Telecommunications Policy*, *32*(7), 490–502.

Grubesic, T.H. (2008b). The spatial distribution of broadband providers in the United States: 1999–2004. *Telecommunications Policy*, *32*(3), 212–233.

Grubesic, T.H. (2012). The US national broadband map: Data limitations and implications. *Telecommunications Policy*, *36*(2), 113–126.

Grubesic, T.H., & Murray, A.T. (2002). Constructing the digital divide: Spatial disparities in broadband access. *Papers in Regional Science*, *81*(2), 197–221.

Grubesic, T.H., & Murray, A.T. (2004). Waiting for broadband: Local competition and the spatial distribution of advanced telecommunication services in the United States. *Growth and Change*, *35*(2), 139–165.

Horrigan, J.B., Stolp, C., & Wilson, R.H. (2006). Broadband utilization in space: Effects of population and economic structure. *The Information Society*, *22*(5), 341–354.

LaRose, R., Gregg, J.L., Strover, S., Straubhaar, J., & Carpenter, S. (2007). Closing the rural broadband gap: Promoting adoption of the Internet in rural America. *Telecommunications Policy*, *31*(6), 359–373.

Liang, B., & Lu, H. (2010). Internet development, censorship, and cyber crimes in China. *Journal of Contemporary Criminal Justice*, *26*(1), 103–120.

Lipsey, R.G., Carlaw, K.I., & Bekar, C.T. (2005). *Economic transformations: General purpose technologies and long-term economic growth*. Oxford University Press.

Mack, E.A., & Grubesic, T.H. (2009). Forecasting broadband provision. *Information Economics and Policy*, *21*(4), 297–311.

Mack, E.A., & Grubesic, T.H. (2014). US broadband policy and the spatio-temporal evolution of broadband markets. *Regional Science Policy & Practice*, *6*(3), 291–308.

Martin, S.P., & Robinson, J.P. (2007). The income digital divide: Trends and predictions for levels of Internet use. *Social Problems*, *54*(1), 1–22.

Monmonier, M. (2014). *How to lie with maps*. University of Chicago Press.

National Telecommunications and Information Administration (NTIA). (2011). *Broadband availability beyond the rural/urban divide*. Retrieved from http://tinyurl.com/qcbbnx3

Norris, P. (2001). *Digital divide: Civic engagement, information poverty, and the Internet worldwide*. Cambridge University Press.

OECD. (2014). *Fixed and wireless broadband subscriptions per 100 inhabitants*. Retrieved from http://tinyurl.com/pm6gqf8

Powell, M. (2010). Broadband keeps the American dream alive & accessible. *Huffington Post*. Retrieved from http://tinyurl.com/lbpevwg

Prieger, J.E. (2003). The supply side of the digital divide: Is there equal availability in the broadband Internet access market? *Economic Inquiry*, *41*(2), 346–363.

Prieger, J.E., & Hu, W.M. (2008). The broadband digital divide and the nexus of race, competition, and quality. *Information Economics and Policy*, *20*(2), 150–167.

Rogers, E.M. (2003). *Diffusion of innovations* (5th Ed.). New York: Free Press.

Strover, S. (2001). Rural internet connectivity. *Telecommunications Policy*, *25*(5), 331–347.

Strover, S. (2003). The prospects for broadband deployment in rural America. *Government Information Quarterly*, *20*(2), 95–106.

Ulanoff, L. (2014). *The national domination of Comcast and Time Warner in 1 map*. Retrieved from http://tinyurl.com/ouxzcu4

Weil, D. (2015). Ex-FCC Chairman Powell: Obama's actions on Net Neutrality 'Shocking'. *NewsMax*. Retrieved from http://tinyurl.com/pz6t3t8

Whitacre, B.E. (2008). Factors influencing the temporal diffusion of broadband adoption: evidence from Oklahoma. *The Annals of Regional Science*, *42*(3), 661–679.

Whitacre, B.E., & Mills, B.F. (2007). Infrastructure and the rural–urban divide in high-speed residential internet access. *International Regional Science Review*, *30*(3), 249–273.

Wu, T. (2003). Network neutrality, broadband discrimination. *Journal of Telecommunications and High Technology Law*, *2*, 141.

Wyatt, E., & Cohen, N. (2014). Comcast and Netflix reach deal on service. *New York Times*. Retrieved from http://tinyurl.com/nuj9l33

2 Broadband policy

Introduction

The twentieth century (1901–2000) was a time of tremendous technological change. It brought innovations such as the television, the personal computer, the mobile telephone and the Internet. Interestingly, the issues surrounding the provision of *access to* and *pricing of* the networks that interconnect these technologies transcend time. At the turn of the century, legislators grappled with a host of new challenges stemming from innovations in communications media. The Communications Act of 1934 was an effort to consolidate these legislative efforts under the purview of one regulatory agency, the Federal Communications Commission (FCC). At this same time, the common carrier status of telephone providers enabled the largest player, AT&T, to function as a regulated monopoly and expand their service footprint to U.S. customers via implicit subsidies.

Until the 1970s, the trend in telecommunications policy was one of increased regulation and oversight. In part, this was fueled by the growing number of communications media, but also by an expanding customer base for these media. The initial actions taken to break up AT&T's monopoly occurred in 1974, moving the industry toward deregulation. In many ways, the Telecommunications Act of 1996 (*96 Act*) characterizes these deregulatory efforts. This law used broad language to promote competition in the private sector with the hope of providing customers with improved service and lower costs. Ironically, the broad language of the *96 Act* created more uncertainty than perhaps intended, which set off a new wave of policy debates that have persisted since the mid-1990s. Thus, in many ways, the current policy debates of today, such as spectrum allocation and universal service, are a blend of policy challenges from both the early and later portions of the twentieth century.

Given the historical underpinnings of broadband policy, this chapter provides a brief overview of telecommunications policy in the United States starting with the Communications Act of 1934 (*34 Act*) and ending with the National Broadband Plan (NBP) introduced in 2010. This includes a discussion of common carriers and the breakup of AT&T, which are critical to understanding the open access and universal service debates of today. This chapter also provides an extended discussion of the Telecommunications Act of 1996 with a focus on the more controversial sections of the legislation, including an overview of the origins of the ongoing municipal broadband debate. Finally, this chapter closes with a brief overview of current federal initiatives to improve access to broadband service and broadband data that include the Broadband Technology Opportunities Program (BTOP), the Broadband Initiatives Program (BIP) and the National Broadband Map (NBM).

Policy arena of the 1930s

This era is not only important from the perspective of telecommunications technology, but also for telecommunications policy. To put things in context, during the 1930s, the telegraph was still a viable means of communication and the radio and telephone were also increasing in popularity. Due to the growing popularity of the radio in the 1930s, this decade is regarded as the golden age of the radio (CHRS, 2014; PBS, 2014). In 1930, 12 million Americans owned a radio, by 1939 ownership had more than doubled to 28 million (PBS, 2014).

Several issues surrounded radio policy at this time. There was high antimonopoly sentiment in the 1930s and it was agreed that government regulation was necessary so that a monopoly did not develop (Benjamin, 1982). People also believed that government regulation was important to allocate radio frequency (Coase, 1959; Moss and Fein, 2003) and that the radio should be managed with the public interest in mind (Benjamin, 1982). Thus, a "market model" was more or less settled upon, but it quickly became clear that this model favored larger broadcasters and did not provide the diversity and equity in programming and content that the public desired (Willihnganz, 1994).

Radio

Many of the debates surrounding radio policy in the 1930s date back to issues from the early days of radio availability. The Radio Act of 1912 (*12 Act*) required that radio operators obtain a license from the Commerce Department. This was an effort to prevent interference and overlap of radio transmissions across the narrow frequency spectrum (Willihnganz, 1994). Although it is a point of debate amongst historians, one interesting hypothesis about the *12 Act* was that it was passed in response to the need for better and timelier communications after the sinking of the *Titanic* (Lasar, 2011a).

By 1924, there were over 1,000 radio stations operating under the regulations of the Commerce Department, which meant that spectrum was unlicensed and free to all broadcasters (Wu, 2007). This lack of spectrum regulation meant that problems of interference were widespread. Broadcasters attempted to circumvent interference issues by operating on specific frequencies or broadcasting at specific times of day to share a particular frequency on the spectrum (Wu, 2007). Although these strategies were able to mitigate some of the interference between broadcasters, the underlying problem of spectrum allocation persisted. The Radio Act of 1927 (*27 Act*), provisions of which would become part of the Communications

Act of 1934 (*34 Act*), was passed to provide some manner of regulating and allocating spectrum to users (Wu, 2007). This Act made two important changes to the operation of radio waves. First, it created a commission that was in charge of splitting the spectrum into different station classes and issuing broadband licenses for radio broadcasts at specific frequencies, times, locations and power levels (Wu, 2007). Second, the *27 Act* prevented censorship of radio broadcasts and ensured all political candidates had equal treatment and equal time on radio broadcasts (Wu, 2007).

Telephone

Around the same time that the radio was growing in popularity, the adoption of a parallel telecommunications medium was also becoming popular, the telephone. In contrast to the radio, which operated on a broadcast (one-to-many) platform, the telephone provided a direct, one-to-one connection between users. First invented in 1876 by Alexander Graham Bell, the Bell System formed nine years later in 1885. Ultimately, AT&T served as aggregator for the Bell System, essentially linking all of the Bell companies together and providing a nearly seamless national coverage (CSUDH, 2014). AT&T did this by purchasing scores of independent telephone companies at the start of the twentieth century (National Research Council, 2002). This was a significant undertaking, as it is important to remember the degree to which the U.S. telephone operators were Balkanized in the nineteenth century. For example, in the 1890s telephone service was relegated to large cities and nearby locales, and AT&T's network remained concentrated in the New England area (Abler, 1977). Between 1890 and 1904, AT&T expanded its service so that it covered nearly two-thirds of the United States (Grubesic, 2003). AT&T also had a competitive advantage over other telecommunications providers because it held patents on long-distance communication (Allen & Koffler, 1999). In 1910, however, the Interstate Commerce Commission began to regulate portions of AT&T's operations to ensure it was providing service to customers at "just and reasonable" rates (ibid.). By the close of the 1930s, over 15 million Americans were connected to AT&T despite declines in long-distance calls during the Great Depression (Telephony Museum, 2014).

Text of the 1934 Act

In 1930, an initial attempt was made to pass what would become the Communications Act of 1934 (MGF, 1935). However, the bill failed because it contained language to minimize the size of government (MGF, 1935). Four years later, on June 19, 1934, the Communications Act of 1934 finally became law and provided a framework for governing the various forms of telecommunications media at that time (MGF, 1935). The simple goal of this law was to consolidate the regulations for radio, television, and telephone and to create a commission (the FCC), which would have oversight over these regulations, plus all interstate and foreign communications (Roosevelt Institute, 2010). In addition, another goal of the *34 Act* was to expand affordable access to communications services (ibid.).

Universal service and AT&T

Perhaps the most enduring legacy of the *34 Act* is the concept of universal service (NCTA, 2013). This legacy is rooted in Section 1 of the Act, which reads as follows:

> For the purpose of regulating interstate and foreign commerce in communication by wire and radio so as to make available, so far as possible, to all the people of the United States, without discrimination on the basis of race, color, religion, national origin, or sex, a rapid, efficient, Nationwide, and world-wide wire and radio communication service with adequate facilities at reasonable charges, for the purpose of the national defense, for the purpose of promoting safety of life and property through the use of wire and radio communication, and for the purpose of securing a more effective execution of this policy by centralizing authority heretofore granted by law to several agencies and by granting additional authority with respect to interstate and foreign commerce in wire and radio communication, there is hereby created a commission to be known as the ''Federal Communications Commission,'' which shall be constituted as hereinafter provided, and which shall execute and enforce the provisions of this Act.
>
> (Section 1, *34 Act*)

Although universal service is a dynamic idea that has evolved over time (Alleman, Rappoport, & Banerjee, 2010), its fundamental goal is a simple one. For example, in the context of broadband, it would mean a broadband connection in every home. This is no different from the idea of universal service as it relates to the telephone. In this context, universal service was defined as "a telephone in every home" (Blackman, 1995, 171). The breakup of AT&T in the 1980s and the pro-deregulation aims of telecommunications policy in the second half of the twentieth century (Wu, 2007) are central to interpreting modern day universal service debates related to broadband. In this sense, the history of AT&T is critical to understanding debates concerning open access and universal service today.

The fact that our current idea of universal service is linked to the telephone is rooted in the service provision strategies employed by AT&T throughout the majority of the twentieth century (Allen & Koffler, 1999; Caristi, 2000). In addition to purchasing independent telephone companies, a second strategy that AT&T used to grow its nationwide network was to cross-subsidize the costs associated with rural telephone provision with the rates (and profit margins) charged to urban and business customers (Caristi, 2000). For this cross-subsidy strategy to work, AT&T had to function as a monopoly, which it was able to do until the formal breakup of the Bell System in 1984. This monopoly power was largely endowed by AT&T's common carrier status, which dates back to the passage of the Mann–Elkins Act of 1910.

From 1910 to the passage of the *34 Act*, both telephone and telegraph providers were considered common carriers according to the Mann–Elkins Act of 1910 and were federally regulated by the Interstate Commerce Commission (ICC; Wu, 2007). A common carrier status meant that in exchange for immunity for any content distributed over the telegraph and telephone networks, carriers had to offer services to any customer willing to pay the prevailing rates, as set by the ICC (ibid.). At this time, AT&T had a monopoly on long-distance service, but did not for local telephone service (ibid.). To compete with these local carriers, AT&T refused to allow local companies to interconnect with their long-distance network, which, in effect, limited the range and value of services provided by local carriers (ibid.). At the close of 1913, AT&T was facing lawsuits in almost every state regarding rates or antitrust issues (Mueller, 1997). On December 19, 1913, AT&T vice-president Nathan C. Kingsbury wrote a letter to the Department of Justice (DOJ) that has become known as the "Kingsbury Commitment" (Mueller, 1997). The Kingsbury Commitment was basically an out-of-court settlement with the DOJ that allowed local carriers to interconnect with the long-distance services provided by AT&T (Wu, 2007). The commitment also allowed AT&T to retain all of the phone companies it had purchased, as well as monopoly control over long-distance services (National Research Council, 2002). However, there were no provisions made for AT&T to connect its local services with other local providers (Mueller, 1997).

The series of lawsuits initiated at the beginning of the twentieth century were just the beginning of many for AT&T. In 1956, AT&T signed a consent decree that settled a 1949 lawsuit that alleged AT&T was monopolizing the purchase of telephone equipment through an exclusive supply relationship by making exclusive equipment purchases solely through Western Electric (Allen & Koffler, 1999). The consent decree allowed AT&T to keep Western Electric but it prohibited them from entering new telecommunications markets, except those pertaining to regulated telecommunications (ibid.).

Although the next major change to national telecommunications policy would not take place until 1996, a lawsuit initiated in 1974 had wide-reaching impacts for the telecommunications industry when it was settled in 1982. This 1974 lawsuit initiated the breakup of the Bell System that had provided Americans with telephone service since the turn of the twentieth century. On January 8, 1982, the Justice Department and AT&T reached an agreement also known as the Modified Final Judgment (MFJ), which required AT&T to unload the local operating companies it had acquired almost a century before (National Research Council, 2002; Wu, 2007). This agreement meant that in 1984, seven regional Bell operating companies (RBOC), Bell Atlantic, NYNEX, BellSouth, Ameritech, USWest, Pacific Telesis and Southwestern Bell would be formed from the consolidation of 22 regional carriers (Wu, 2007). These seven companies, also known as "Baby Bells," were to be tasked with providing connectivity within their service areas while AT&T retained the rights for providing all long-distance service for the Baby Bells and the right to manufacture telecommunications equipment (ibid.). AT&T could not be a stockholder in any of the RBOCs and the RBOCs were not allowed to manufacture telecommunications equipment (ibid.).

This 1984 breakup of the Bell System is important to understanding the outcomes of interconnection provisions in the *96 Act* enacted twelve years later and associated efforts to circumvent these provisions by incumbent local exchange carriers (ILECs). The common carrier status enjoyed by AT&T since the Mann–Elkins Act of 1910 is also important to understanding the disparate status of broadband service provided over telephone lines, as opposed to broadband services provided over cable lines, which remains an ongoing source of controversy to this day in the form of the open access debate.

Policy arena of the 1990s

Although the core infrastructure of the Internet (i.e. ARPANET) was initially developed in the mid-1960s (Abbate, 1999), it was not until 1995 that the federally funded and developed backbone of the Internet, including NSFNET, was transferred into the care of private telecommunications providers (Abbate, 1999). This commercialized version of the Internet backbone enabled customers to gain access to the Internet and complementary services by internet service providers (ISPs). From this backbone, four types of ISPs are involved in the provision of service: transit backbone ISPs, downstream ISPs, online service providers and web hosting firms (Cukier, 1998). The bulk of customers interact with downstream ISPs, online service providers and web hosting firms. Downstream ISPs consist of local and regional providers that serve individuals and businesses such as Comcast and AT&T by connecting them to the Internet backbone (Gorman & Malecki, 2000). Web hosting companies host websites so that internet users can access them (ibid.); examples of web hosting companies include Startlogic, Dreamhost and Bluehost. The top layer of ISPs is online service providers (OSPs), all of which are highly recognizable to consumers. OSPs include companies such as Ebay, Amazon, Google and America Online.

Between 1992 and 1998, most customers accessed the Internet with dial-up connections (Grubesic, 2006). Dial-up connections required internet users to dial into the network on a per use basis. Starting in the late 1990s customers began to gain high-speed access to the Internet (ibid.). In addition to faster speeds offered by these broadband connections, the fundamental service model also differs from dial-up or narrowband connections because it is considered an "always-on" internet connection that does not require users to dial into the network each time they wish to use it (FCC, 2015a). In the late 1990s there were three primary industries competing in broadband markets: cable, telephone and fixed wireless (Grubesic & Murray, 2004). Aside from the relative uncertainty about which platform would eventually dominate, there were also significant challenges with providing "last-mile" service to customers. These challenges stemmed from the high fixed costs associated with deploying last-mile infrastructure and the uncertainty in recouping these costs. Thus, providers preferred (and still do) to deploy infrastructure in densely populated, high-income, urban areas to maximize their chances of recouping these enormous fixed costs (Kolko, 2012). This cost structure to telecommunications provision and the low levels of competition within the local loop (Grubesic, 2003) meant that many customers did not necessarily have access to broadband infrastructure in the late 1990s and early 2000s. Since cross-subsidies from urban customers could not be used to subsidize rural customers at the time the *96 Act* was formulated and passed, access to the last mile of broadband represented a tremendous deployment and regulatory challenge, (National Research Council, 2002). The impacts of the *96 Act* are discussed more thoroughly in the next section.

The early 1990s also represented a time of tremendous activity and upheaval in the telecommunications market. In addition to the development of the World Wide Web, web browsing software (e.g., Mosaic), and commercial email (e.g., Hotmail), platforms were also making headway into the OSP market (Lanxon, 2008). It was also a period of tremendous merger and acquisition activity in the telecommunications market (Warf, 2003). Qwest acquired USWest; NYNEX merged with Bell Atlantic, which then merged with GTE to form Verizon; SBC acquired Pacific Telesis and Ameritech; MCI merged with both WorldCom and then UUNet; and AT&T acquired TCI and several other cable companies (National Research Council, 2002). In many ways, these mergers and acquisitions set the stage for the next 20 years of wireline and wireless telecommunications provision.

The Telecommunications Act of 1996

The passage of the Telecommunications Act of 1996 represented the first wholesale change in U.S. telecommunications policy since the Communications Act of 1934. The *96 Act* also reflects the pro-deregulation aims of telecommunications policy that have shaped the second half of the twentieth century (Wu, 2007). The goal of the Act was to deregulate the telecommunications industry and provide consumers with the benefits of higher quality services and lower prices afforded by competitive forces (*96 Act*). Policymakers also hoped this groundbreaking legislation would facilitate widespread deployment of "new telecommunications technologies." To encourage competition, the *96 Act* is founded upon unbundling and facilities-based competition (National Research Council, 2002). Unbundling provisions are aimed at ILECs, while facilities-based competition revolves around competitive local exchange carriers (CLECs). CLECs are essentially new telecommunications entrants that were looking to use the existing infrastructure built by ILECs to provide service and compete within local markets (ibid.).

Despite the pro-deregulation and pro-competition aims of the *96 Act*, it has been the source of much controversy. Opponents were concerned that the goal of market competition would hurt rural areas by removing any chances for cross-subsidies to improve access and service quality for rural customers (Compaine & Weintraub, 1997). Another issue with the *96 Act* is its lack of specificity, particularly regarding which actors should be involved in providing broadband services, how the services would be utilized and the manner in which broadband platforms and services would be deployed (Grubesic & Murray, 2004). This ambiguity, for better or worse, has influenced nearly all of the regulatory and economic conditions of the current broadband ecosystem. In fact, some experts have suggested that the *96 Act* is nothing more than "a public policy experiment in the U.S., testing the viability of vertical de-integration, for example through the leasing of unbundled network elements" (Clark, Gillett, Lehr, Sirbu, & Fountain, 2003). Table 2.1 provides an overview of some of the more salient and controversial aspects of the *96 Act*, all of which have provided fertile ground for a continued onslaught of legislative efforts to combat the outlined provisions. To provide some perspective on these issues, the next several subsections of this chapter review and describe the most controversial components of the *96 Act* and their implications for broadband.

Section 251 and unbundling

The unbundling provision of the *96 Act* is perhaps one of the more controversial sections of the legislation. Section 251 requires that ILECs make portions of their networks available to CLECs on "a just, reasonable and non-discriminatory basis" (96 Act). In essence, this portion of the *96 Act* is an effort to correct what had become a bottleneck in internet traffic by allowing more providers access to local loop facilities (Lehr & Glassman, 2001). Local loops are the "on" and "off" ramps to the Internet and ILECs spent billions of dollars, albeit with the help of subsidies, on the infrastructure that permits information ingress and egress to the system (ibid.). In short, this is infrastructure that most CLECs cannot afford to recreate because of their smaller size, access to capital and local focus. Thus, an arrangement where ILECs leased the infrastructure to CLECs provided a means of making costly local loop facilities available in the hopes of fostering competition in the provision of advanced telecommunications services (ibid.). After passage of the *96 Act* however, it has been argued that ILECs engaged in anticompetitive behavior in several ways, including refusals to lease telephone lines, late completion of local loop installations and predatory pricing (Kuschnick, 2001). In fact, studies estimate that since the passage of the *96 Act*, over \$2 billion worth of fines have been assessed to ILECs for actions that are in violation of the *96 Act* (Freese, 2002).

In 2001, legislators proposed the Tauzin-Dingell Bill (H.R. 1542), also known as the Internet Freedom and Broadband Deployment Act (Lehr & Glassman, 2001; Thierer, 2001). The goal of the bill was to exempt ILECs from the unbundling provisions of the *96 Act* to make it easier for them to offer long-distance internet service (Ferguson, 2002). The ILECs who supported this bill argued that competition in broadband markets would not suffer because competition already existed through cable companies who provided the lion's share of broadband service at this time (ibid.). Opponents, however, asserted that this legislation would reduce ILECs' incentives to make investments in broadband facilities, and would force CLECs to build their own last-mile infrastructure (e.g., switching centers

Table 2.1 State regulations for municipal broadband, 2015

State	Type Regulation	Statute	Details
Alabama	Restricted	Alabama Code § 11-50B-1 et seq.	Municipalities must comply with a variety of restrictions. For example, municipalities must demonstrate that communications services such as telephone, Internet and television, are individually self-sustaining. Municipalities wishing to offer cable services must successfully pass a referendum before offering such services.
Arkansas	Prohibited	Ark. Code § 23-17-409; Act 1050	Government entities may not provide local exchange services, data, broadband, video and wireless. Municipalities that have their own electric and cable utilities are excluded from this prohibition and may offer telecommunication services but not basic exchange services such as telephone service.
California	Restricted	California Government Code § 61100 (af)	Community Service Districts (CSD) may only provide communications services as long as no private entity is willing to do so. If the CSD builds the network and a private party emerges later and is willing to provide such services, the CSD must sell the entity its network at "fair market value."
Colorado	Regulated	Colo. Rev. Stat. Ann. § 29-27- 201 et seq.	Municipalities must have a referendum to provide services unless the community is unserved and incumbent providers refuse to provide service.
Florida	Regulated	Florida Statutes § 350.81	Municipalities must demonstrate that their network will break even within the first 4 years of operation. Municipal telecommunications services are taxed.
Louisiana	Regulated	La. Rev. Stat. Ann. § 45:884.41 et seq.	A referendum is required before construction and operation of networks can begin. Costs of operating a similar network by private providers must be imputed. Any obligations faced by incumbent provider franchisees (franchise fees, institutional networks) are void until community networks can satisfy a variety of benchmarks.
Michigan	Regulated	Mich. Comp. Laws Ann. § 484.2252	Communities may only build telecommunications networks if fewer than 3 qualified bids are received after issuing a request for proposals (RFP). If the community builds the network, they must comply with the guidelines outlined in the original RFP.
Minnesota	Regulated	Minn. Stat. Ann. § 237.19	A super majority referendum of 65% is needed for a municipality to offer telecommunications services.

State	Status	Statute	Description
Missouri	Prohibited	Mo. Rev. Stat. § 392.410(7) and Mo. Rev. Stat. § 71.970	Cities and counties may offer cable services after a referendum is passed. They may not offer telecommunications services. There are exceptions for telecommunications services used for internal, educational, emergency, health care, and "Internet-type" services.
Montana	Restricted	Mon. Code Ann. § 2-17-603	A city or town can provide Internet services as long as these services started prior to July 1, 2001 or Internet service is not available within the jurisdiction of the city or town interested in providing such services. If a provider does step forward at a later date, the provider must give the city or town 30 days' notice prior to offering Internet services. Similarly, the city or town must notify current subscribers of the presence of this private entity within 30 days of receiving notice from the private entity. That city or town can also decide to stop providing Internet services within 180 days after notifying current subscribers of the presence of the private entity.
Nebraska	Prohibited	Neb. Rev. Stat. Ann. § 86-594 and Neb. Rev. Stat. Ann. § 86-575	Communities and public power companies may not wholesale or retail offer telecommunications services which includes wholesale or retail broadband, Internet, telecommunications, or cable services. Public power utilities cannot provide retail services but they can sell or lease dark fiber as wholesalers at rates lower than those charged by incumbents in the same market.
Nevada	Prohibited	Nevada Statutes § 268.086 and Nevada Statutes § 710.147	Uses a population threshold as basis for service prohibition. Municipalities with more than 25,000 people and counties with greater than 50,000 people may not offer telecommunications services.
North Carolina	Restricted	Session Law 2011-85, House Bill 129	Unless a city network existed previously and was grandfathered in before the passage of this statute, a city cannot provide communications services without complying with numerous requirements. For example, public providers must hold a referendum before service can be provided. They are also required to make sensitive information available to private companies and must charge higher costs to account for costs that private providers might have to pay if similar service were provided. Several public financing options are also not available for use in obtaining funds to build the network.

(continued)

Table 2.1 (continued)

State	Type Regulation	Statute	Details
Pennsylvania	Restricted	66 Pa. Cons. Stat. Ann. § 3014(h)	Communities can only provide service if the local telephone company refuses to provide services at speeds requested. Price is not a factor in the provision of service by local telephone companies, speed is the relevant factor. If the local telephone company is willing to provide services at the speeds requested no other factor can be considered including price, quality of service, etc.
South Carolina	Regulated	S.C. Code Ann. § 58-9-2600 et seq.	The state places a variety of restrictions and regulations on communities that want to invest in and build their own networks. This includes a referendum before offering services and complying with all legal requirements that apply to private providers. Communities must also estimate or impute the costs that private companies would otherwise pay to build a comparable network. Items that must be included in these imputed costs include taxes, subsidies and government stimulus funds.
Tennessee	Regulated	Tennessee Code Ann. § 7-52- 601 et seq	Regulations split municipalities into two groups, municipalities that own electric utilities and those that do not. Municipalities that do own an electric utility may provide telecommunications services but must comply with a range of disclosure, hearing, and voting requirements amongst others. Municipalities that do not own electric utilities are restricted to the provision of telecommunications services in "historically unserved areas." To provide services in these areas, municipalities must offer service via joint ventures with private sector entities.
Texas	Prohibited	Texas Utilities Code, § 54.201 et seq.	Municipalities and public utilities may not offer telecommunications services.

State	Status	Code	Description
Utah	Restricted	Utah Code Ann. § 10-18-201 et seq.	Retail municipal providers must conduct feasibility studies to verify that the network will generate positive cash flows in year 1 of services provision and for 5-year projections. If separate telecommunications services (cable, Internet) are provided, it must be demonstrated that each of the services will generate positive cash flows separately.
Virginia	Regulated	VA Code § 15.2-2108.6, VA Code § 56-265.4:456-484.7:1, and VA Code § 56- 484.7:1	Municipal electric utilities (MEUs) may offer telecommunications services subject to a variety of reporting requirements. If MEUs want to offer cable services, they must demonstrate that their network will generate positive cash flows within the first year.
Washington	Restricted	Wash. Rev. Code Ann. § 54.16.330	Municipalities may act as a wholesaler for telecommunication services, but may not provide telecommunications services directly to customers.
Wisconsin	Regulated	Wis. Stat. Ann. § 66.0422	Cities and towns wanting to provide telecommunications, cable, or Internet services must conduct feasibility studies and hold public hearings before doing so. Subsidies are prohibited for most cable and telecommunications services and minimum prices prescribed for telecommunications services.

Source: CyberTelecom (2013), CBN (2015), Brodkin (2014b), Baller Herbst Law Group (2014)

and wires; Grubesic, 2003). Although the bill never became law, in February 2003, the FCC rescinded portions of the *96 Act* that classified telephone lines as network elements (Phillips & Tessler, 2003). This meant that CLECs would need to negotiate the leasing of telephone lines with incumbents, but that incumbents would not have to share newly constructed network elements with competitive local exchange carriers (ibid.).

Section 303 and the open access debate

Section 303 is controversial because it did not require cable companies to obtain franchises to provide broadband services. The removal of franchise restrictions for cable providers highlights another instance in which cable broadband providers have received more favorable treatment than other broadband providers that do not have a "common carrier" status. A common carrier offers communications service to the public for a fee but does not control the content of information transmitted (National Research Council, 2002). As noted previously, the common carrier status of telephone companies dates back to the Mann–Elkins Act of 1910 and means that any internet services offered over telephone lines are subject to common carrier regulations (Wu, 2007). Cable providers were never classified as common carriers, however, which means they were never required to offer service on a non-discriminatory basis and they were not required to allow unaffiliated content providers access to their distribution systems (Speta, 2000).

Compared to common carriers, cable providers have always faced somewhat limited regulation (Crandall, Sidak, & Singer, 2002). Under the Communications Act of 1934, local television signals transmitted over cable lines were subject to basic rate regulations (National Research Council, 2002). Cable companies are also regulated locally via franchise agreements with municipal, county or state authorities (Caristi, 2000; Gillet, Lehr, & Osorio, 2004), which also serve as a source of revenue for municipalities (National Research Council, 2002). Unfortunately, franchise agreements do not require cable companies to upgrade their networks, and if upgrades are made for the purposes of offering broadband services, they are not required to provide access to unaffiliated providers (ibid.). In short, the language of the *96 Act* removed the somewhat limited regulation for cable companies via franchise agreements at the municipal level and created a further divide in the level of regulation facing common carriers when compared to cable companies.

These differences in regulation are the crux of the local access debate. Proponents of open access argue that when cable companies provide access to the Internet, they function as common carriers and should be subject to the same access obligations as common carriers (Picot & Wernick, 2007). Cable companies argue, however, that when they provide internet service they are providing a cable service rather than a telecommunications service and should remain exempt from common carriers obligations (National Research Council, 2002). Until very recently, little was done to resolve this open debate. In 2002, Senator John Breaux of Louisiana introduced the Broadband Regulatory Parity Act of 2002

(S. 2430). Section 3 of the 2002 Parity Act amended the Communications Act of 1934 (47 U.S.C. 251 et seq.) to ensure equal regulatory treatment of all providers of broadband service (S. 2430). In response, the FCC reclassified broadband as an "information service" rather than a telecommunications service (National Research Council, 2002; Miller, 2014). This meant that providers offering information services were not regulated under common carriage provisions; this was the case for both cable providers and ultimately DSL providers (Wu, 2007). This reclassification of broadband as an information service was later upheld in the U.S. Supreme Court case of *National Cable and Telecommunications Association vs. Brand X Internet Services*. The Supreme Court affirmed the FCC's decision to classify cable providers of broadband as providing information services (Hansell, 2005). Of note, however, is that on February 26, 2015, with a 3 to 2 vote, the FCC approved net neutrality rules and classified broadband service as a utility (Somer, 2015).

Section 706 advanced telecommunications

The main problem with Section 706 of the *96 Act* is the ambiguous definition of advanced telecommunications capabilities:

> The term 'advanced telecommunications capability' is defined, without regard to any transmission media or technology, as high-speed, switched, broadband telecommunications capability that enables users to originate and receive high-quality voice, data, graphics, and video telecommunications using any technology.

Clearly, the lack of specificity in this definition provides little information to delineate broadband in terms of upload and download speeds or the platforms over which broadband is to be delivered (Grubesic & Murray, 2004). While this likely serves as insulation and/or protection against the evolutionary nature of broadband, the indistinctness in this section of the *96 Act* provides little incentive for companies to make substantial investments in a particular platform. In part, this is because there was significant uncertainty regarding which broadband platform would continue to appeal to consumers or dominate markets in the future. Thus, if providers locked-in to a particular technology, they could face greater potential losses.

Section 254 and universal service

It has been stated that one of the goals of the *96 Act* was to move away from *implicit* subsidies toward *explicit* subsidies for broadband service (Allen & Koffler, 1999). An implicit subsidy happens when a company obtains revenues from some source at rates that are above cost, and then prices services for other sources at rates below cost (FCC, 1997). In other words, implicit subsidies are strategies for subsidizing service for some users by overcharging other users for service. One example of an implicit subsidy is geographic rate averaging which involves charging urban consumers higher rates than rural consumers (Allen & Koffler, 1999). A second example of an implicit subsidy is the use of higher rates for businesses to subsidize lower rate structures for residential consumers (ibid.). Alternatively, explicit subsidies are specific and visible budgetary outlays made by the government (Valdés, 1988).

This issue of subsidies is important because it is related to the manner in which universal service is provided. Historically, AT&T used implicit subsidies to provide universal telephone service by charging urban customers more for service than rural customers (Caristi, 2000). While people believe that the conceptual foundation for universal service is rooted in the preamble of the Communications Act of 1934, the Congressional record does not actually contain any conversation about universal service (Mueller, 1997). Where the *96 Act* is concerned, Section 254, specifically 254(b), outlines several key tenets for universal service (Allen & Koffler, 1999). These tenets include language mandating that quality service should be available to all, at just, reasonable and affordable rates (ibid.).

The origins of the universal service debate and the potential mechanisms for achieving universal service via regulated monopolies and rate subsidies can be traced back to the rollout of telephone service at the close of the nineteenth century (Mueller, 1993; Blackman, 1995). In fact, Theodore Vail, founder of the Bell System, first used the phrase "universal service" in 1907 (Blackman, 1995). As discussed previously, the current concept of universal service is strongly tied to the manner in which AT&T provided telephone service in the first half of the twentieth century (Allen & Koffler, 1999; Caristi, 2000). In the early days of the telephone system, universal service was accomplished via rate averaging, which allowed long-distance service to subsidize local service (Gasman, 1998). Some argue, however, that the manner in which Vail used universal service referred to the idea of interconnection. This is not how universal service is thought of today (ibid.). In fact, Mueller (1997) suggests that the concept itself is generational, evolving over time.

The first generation of universal service spans the years 1907–1921. Its primary goal was to unify the network so that all telephone subscribers could communicate with one another (ibid.). The second generation of universal service was primarily focused on technology diffusion and ensuring that there was a telephone in every home. Finally, the third generation deals with the provision of service in an era of globalization where there are multiple service standards to contend with (ibid.).

Debate about the second-generation idea of universal service began in the 1950s with conversations about separations principles (ibid.). This debate revolved around whether setting rates via the board-to-board method or the station-to-station method was better for serving customers (ibid.). The station-to-station method uses long-distance revenues to subsidize local service while the board-to-board method uses higher local rates to subsidize long-distance rates. Over the years, debates over the separation of rates resulted in an effective rate-setting scheme based on the station-to-station method (ibid.). Although this concept worked well when AT&T functioned as a regulated monopoly, this service model began to break down in the 1970s as competition in the long-distance market heightened and new entrants to the long-distance market, including Sprint and MCI, were able to obtain local business lines from AT&T at subsidized rates. This provided Sprint and MCI with a significant cost advantage (ibid.).

After the breakup of AT&T in the 1980s, a combination of growing competition in the telecommunications industry and the passage of the Telecommunications Act of 1996 prompted funding for universal service to be moved from implicit subsidies via station-to-station rate separation agreements to explicit subsidies in the form of universal service funds from the federal government (Dippon, Huther, & Troy, 2010). Proponents of universal service argue that funds are necessary because the service is essential and because there are network effects associated with telecommunications use (Rosston & Wimmer, 2000). Network effects mean that the value of a product is derived from the number of people using that product (Varian & Shapiro, 1999). As alluded to previously, the *96 Act* mandated that the concept of universal service be updated and funding mechanisms devised for the provision of advanced telecommunications capabilities for all (Allen & Koffler, 1999). Thus, although the discussion surrounding universal service after passage of the *96 Act* centered on the creation and allocation of monies to a universal service fund (Gasman, 1998), there were four additional issues worth highlighting (Mueller, 1997, 165):

1   The cost of universal service obligations for incumbent telephone companies.
2   Mechanisms by which universal service costs could be financed within a competitive environment.
3   The technical and price arrangements required to interconnect incumbent providers with new competing networks.
4   Options for revising the concept of universal service to account for new technologies.

Both the Universal Service Administrative Company (USAC) and a Federal State Joint Board on Universal Service were created after the *96 Act* passed (FCC, 2015b). The Universal Service Fund (USF)[1] is funded by inter-state and internation end-user revenues collected from telecommunications providers (wireline and wireless companies, interconnected Voice over Internet Protocol (VoIP) providers and cable companies providing voice services; FCC, 2015b). Four programs are supported by the USF: the High Cost Program, the E-rate Program, the Low-Income Program and the Rural Health Care Program (Dippon et al., 2010; FCC, 2010; FCC, 2015b). The High Cost Program subsidizes telecommunications services in areas where the costs to obtain service are prohibitively high otherwise (i.e., remote rural areas), while the E-rate Program subsidizes voice and internet connections for schools and libraries (FCC, 2010). Of these programs, the E-rate Program has proven particularly successful for providing schools and libraries with internet access. At the time the *96 Act* was passed, only 14% of K-12 classrooms nationwide had access to the Internet (FCC, 2015b). Today, nearly all schools have internet access (FCC, 2015b). In an effort to expand and modernize the reach of this program to cover broadband services, the FCC passed the E-Rate Modernization Order and the Second E-rate Modernization Order in 2014 (FCC, 2015b).

It is worth noting that the USF could play a critical role in achieving universal broadband service for customers in rural areas. It has been suggested that the expansion of this fund could provide the stimulus needed for extending broadband services to rural customers and subsequently meeting the goals of the *96 Act* (Parker, 2000). The current USF was designed to support schools, libraries, and rural healthcare, but not broadband directly (Parker, 2000; FCC, 2010). Thus, current policy discussions about universal service are focused around mechanisms for translating support for telephone networks to broadband networks (FCC, 2010).

The NBP (FCC, 2010) suggests a three-stage roadmap for moving universal service into the broadband age. This roadmap involves the creation and funding of a Connect America Fund (CAF), a Mobility Fund and the elimination of High Cost Programs from the existing USF docket (ibid.). In particular, the CAF, which would support broadband deployment in unserved areas, would be created with money saved from cost-cutting measures implemented for the existing USF (Dippon et al., 2010). Further, the goal of the Mobility Fund is to provide initial, one-time only, financial support for the deployment of advanced wireless networks in remote areas such as Alaska and West Virginia, both of which suffer from low coverage rates (FCC, 2010). These funds would not only facilitate the deployment of more cell sites for 4G networks in these areas but also benefit public safety agencies that use commercial telecommunications services (ibid.).

Aside from the expansion of universal service funding to cover gaps in broadband and mobile telecommunications, universal service debates are increasingly focused on questions related to broadband use and the reasons that customers do not subscribe to broadband services (Levin, 2010). As Levin (2010) notes, there are two main reasons that customers may not use broadband service. First, they do not have a use for the service. Second, they cannot afford the service (ibid.). While a policy intervention is not necessary for customers that do not perceive a need for service, one may be required for prospective customers that cannot afford it. A close inspection of the NBP (FCC, 2010) suggests that policy measures are under development for improving broadband access, affordability and the digital literacy of potential broadband users. Research efforts include data collection to deepen our understanding of the characteristics of non-adopters and to find ways around barriers to non-adoption such as a perceived lack of relevance, cost, and disabilities (ibid.). In this context, it is important to note that the effectiveness of developed policies for mitigating broadband gaps is strongly related to data quality and uncertainty. The problems and prospects of using broadband data from the NBM and other sources are discussed at length in Chapter 4 of this book.

Section 253a and municipal broadband

Aside from the use of federally controlled universal service funds to ameliorate disparities in broadband availability, state and local broadband deployment efforts are also powerful tools to combat disparities in last-mile access (National Research

Council, 2002). However, the involvement of municipalities in the provision of broadband service is controversial. Proponents of municipal broadband see it as a viable strategy for overcoming a market failure to provide communities with acceptable levels of quality broadband service. Conversely, opponents suggest that municipal efforts provide competition to the private sector, ultimately discouraging investment by private companies (Spiwak, 2015). Further, municipal efforts can strain limited public resources and there is very little evidence identifying pathways to profitability for cities making such investments (Caristi, 2000; Spiwak, 2015). In particular, what happens when networks fail? This ongoing controversy over municipal broadband can be linked to the verbiage of Section 253a of the Telecommunications Act of 1996 (Úrzua 2004, 42). This section of the Act reads as follows:

> No State or local statute or regulation, or other State or local legal requirement, may prohibit or have the effect of prohibiting the ability of any entity to provide any interstate or intrastate telecommunications service.

In this section, there is ambiguity associated with the word "any." As a result, some states have passed laws to clarify whether or not municipalities are permitted to deploy broadband (Úrzua, 2004, 46). In fact, some states have even used policy to prohibit municipalities from involvement in offering and/or building broadband networks and associated services.

These state efforts are rooted in a U.S. Supreme Court battle that began in Missouri after the passage of a state law prohibiting the provision of broadband service by municipalities. Specifically, in 1997, Missouri municipalities, municipally owned utilities and other municipal-level organizations banded together to form a "municipal league" to contest a state law that prohibited "political subdivisions" of the state from selling telecommunications services or facilities to ISPs (either public or private; Stricker, 2013). Based on the language of Section 253a of the *96 Act*, which states the FCC can "preempt the enforcement of such a statute . . . to the extent necessary to correct such violation or inconsistency" (ibid.), the municipal league submitted a petition to the FCC that asked it to declare that such a law "preempted" Statute 253a of the *96 Act* (ibid.). Ultimately, although the FCC agreed that the law did not correspond with the spirit or goals of the 1996 Act, the FCC did not believe it preempted the Missouri law as requested because the phrase "any entity" in the *96 Act* was not meant to include political subdivisions (ibid.).

The law eventually reached the U.S. Supreme Court in March 2004, where the Court overturned the Eighth Circuit Court decision in *Nixon vs. Missouri Municipal League* and determined that the Missouri law was valid (ibid.). This decision effectively gave states the right to restrict municipalities from building broadband networks and prevent them from offering broadband service. The decision was, and continues to be, an important landmark in telecommunications history. Almost immediately after the Supreme Court decision, 13 states passed legislation related to the involvement of municipalities in the provision of broadband service (Gillett, 2005). Some states prohibited any activity in this sector, while other states placed restrictions on the activities of municipalities that range from financial considerations to the type of business model (wholesale vs. retail) they are allowed to utilize (Gillett, 2005, 585). Today, 21 states have restrictions on the involvement of municipalities in the provision of broadband infrastructure and services (CBN, 2015).

Figure 2.1 highlights the states with municipal regulations on broadband and divides them into four categories based on the strength of their oversight: 1) Not Regulated, 2) Regulated, 3) Restricted and 4) Prohibited. Table 2.1 provides additional detail about the restrictions in each of these states. The most common type of oversight on municipal broadband is some type of regulation. This can include the ability to demonstrate that a municipal network will be self-sustaining, as mandated in Alabama, to a super majority referendum of votes to offer telecommunications services, which is the case in Minnesota. Perhaps the most common restriction is the mandate that there be a referendum before broadband services can be offered. Frequently, the need for a referendum is combined with other requirements, such as the estimation of costs of what it would take for a private company to build a broadband network, as is required in South Carolina. Table 2.1 also highlights that five states have placed a prohibition on municipal broadband: Arkansas, Missouri, Nebraska, Nevada and Texas. While the Nevada legislation may not be considered a complete prohibition because of its use of a population threshold of 25,000 to limit service, the use of this population threshold is actually quite restrictive. Of the 19 municipalities in the state of Nevada, 13 have populations of 24,999 or less (National League of Cities, 2013). Thus, 68% of these locales cannot offer municipal broadband services. In short, although some of the state restrictions seem innocuous on the surface, they are actually more restrictive than they appear.

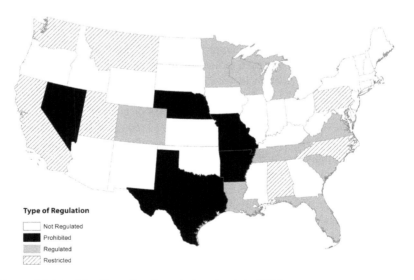

Figure 2.1 Municipal broadband regulations by state, 2014.

Perhaps the most interesting trend to this type of legislation is the growing comprehensiveness of its proposed limitations. For example, a 2014 bill submitted in Kansas on behalf of cable companies sought to prevent municipalities from offering video, telecommunications or broadband, and from purchasing, leasing, maintaining, operating or building the infrastructure or facilities for providing video, telecommunications or broadband services (Brodkin, 2014a). This bill is important because it effectively prevents any public–private partnerships designed to assist in broadband deployment efforts, such as the Google Fiber initiative, which brought fiber broadband to Kansas City (ibid.). The battle over the ability of the FCC to preempt state laws prohibiting municipal broadband continues to this day. President Obama has asked the FCC to preempt state laws that place limitations on municipal broadband projects (Gross, 2015). North Carolina and Tennessee are battleground states for these types of changes (ibid.). The status of this preemption effort has yet to be determined. Ironically, the arguments in favor and against municipal efforts remain largely unchanged from those following the passage of the *96 Act*. For more details on municipal efforts and the best practices for rolling out municipal broadband, see Chapter 8 of this book.

Although many roadblocks remain for rolling out broadband infrastructure – some legal, some technical and others related to geography – innovative private sector efforts to improve broadband access are growing. As mentioned earlier, in 2011, Google entered the ISP market by announcing that fiber-to-the-home would be rolled out to consumers in Kansas City, MO and Kansas City, KS at gigabit speeds (Matteson, 2013; Canon, 2014). Google Fiber is currently available in three cities: Kansas City, MO/KS, Provo, UT and Austin, TX (Kish, 2015). On January 27, 2015, Atlanta, Charlotte, Nashville and Raleigh–Durham were announced as the four newest Google Fiber cities (ibid.). Other cities that may become Google Fiber cities in the near future include San Jose, CA, Portland, OR, Salt Lake City, UT and San Antonio, TX (ibid.).

Not only do Google initiatives represent a promising alternative to traditional private ISPs and municipal-based efforts, there is evidence that these initiatives have an impact on business activity. The Kansas City Startup Village, which began in 2012, provides some evidence of these business impacts (KCSV, 2015). It is an entrepreneur-led community whose mission is to educate and support entrepreneurs through programming and knowledge sharing (ibid.). While several studies have noted that broadband is not a "cure all" for economies in remote areas (Hales, Gieseke, & Vargas-Chanas, 2000; Mack, 2014) these studies, combined with the recent evidence from Kansas City, suggest that broadband is an important component to encouraging new business activity, which is vital to the continued development of regional economies.

Recent federal broadband initiatives

Aside from community level and private sector efforts, initiatives at the federal level to improve broadband access continue. As discussed earlier, the NBP plays an important role in this domain. After two years of information collection that involved 36 workshops, 31 public notices, 1,100 ex parte filings and hundreds of thousands of page content and comments, the NBP was released to the public in March 2010 (FCC, 2010). This plan is a strategic agenda for deploying broadband infrastructure and enhancing the adoption of broadband applications for use by individuals and multiple sectors of the economy including health care, education, energy, government and civic engagement (Grubesic, 2012). The plan also outlines an agenda for dealing with persistent issues that date back to the early years of telecommunications regulation such as licensing of the wireless spectrum and universal service. An important aspect of this plan was the creation of an NBM using federal funds to improve the data available for researchers and policymakers (FCC, 2010; Grubesic, 2012).

In 2009, Congress also passed the American Recovery and Reinvestment Act (ARRA). As part of this Act, $7.2 billion dollars was provided to both the Rural Utilities Service (RUS) of the U.S. Department of Agriculture and the National Telecommunications and Information Administration (NTIA; NTIA, 2015). Of these monies, $4.7 billion dollars were allocated to the NTIA for the BTOP and $2.5 billion dollars were award to the RUS for the BIP (Kruger, 2012). The goals of the BTOP program are threefold: 1) to encourage broadband deployment, 2) to improve public computer centers and 3) to encourage broadband adoption (NTIA, 2015). The ARRA also directed the RUS to allocate funds through the BIP to rural areas without sufficient access to broadband for economic development purposes (Kruger, 2012). An emphasis of these programs is that they allocate funds to projects that would facilitate the deployment of broadband in unserved and underserved areas (Kruger, 2012). Unserved areas are defined as locales where 90% of households are without access to terrestrial broadband (ibid.). Underserved areas are those that meet at least one of following criteria (ibid.):

1   "no more than 50% of the households in the proposed funded service area have access to facilities-based, terrestrial broadband service at greater than the minimum broadband transmission speed";
2   "no broadband service provider advertises broadband transmission speeds of at least 3 megabits per second (Mbps) downstream"; or
3   the rate of broadband subscribership for the proposed funded service area is 40% of households or less.

In 2009, the State Broadband Data and Development (SBDD) program was launched with two purposes in mind. The first purpose was to enhance the use of broadband for improving the competitive position of state and local economies. The second purpose was to assist in the collection of data, on a bi-annual basis, on the availability, speed and location of broadband service; as well as data collection on broadband usage by schools, libraries and hospitals (NTIA, 2015). Data collected by the SBDD, in coordination with the NTIA, is the source of the information in the NBM, which was completed on February 17, 2011 (ibid.). The release of the NBM, which critics have dubbed a "map to nowhere," was an effort to make broadband data more readily available given issues with data available from the FCC in the Form 477 database (Lasar, 2011b). While the availability of broadband data at the Census block level is a marked improvement over ZIP code and Census tract level data previously available from the FCC, there are several nuances to these data that are important to be aware of for spatial and economic analyses (Grubesic, 2012). The map contains no pricing information and relies on self-reported speed data from providers (Lasar, 2011b). In addition, analyses of the NBM data have uncovered variations in provider participation by state and overestimates of broadband coverage (Grubesic, 2012). The problems and prospects of these data, as well as important nuances for spatial analyses of these data are discussed in Chapter 4.

Conclusions

The flurry of activity at the federal, state and local level concerning broadband highlights the high level of importance that the United States places on emerging telecommunications technologies and their role in day-to-day activities, research, the economy and regional development. In many ways, the NBP highlights the core issues pertaining to the digital divide that have emerged in policy development circles over the past few decades. However, broadband access is no longer a divide between the "haves" and "have-nots"; instead, it is a more nuanced, multifaceted issue that covers issues of provision, cost of access, platform and provider choice, speed and quality of service (Grubesic, 2015). The classification of areas into unserved and underserved areas for the allocation of BTOP and BIP funds is an indication of the need for this more spatially informed approach to addressing the digital divide, as are the speed improvement initiatives in the NBP (FCC, 2010).

The overview of twentieth and twenty-first century policy initiatives in this chapter highlighted the nuanced and historical underpinnings of telecommunications policy, which are critical to understanding present policy efforts to close the broadband divide. This recent incarnation of the digital divide will likely require that a host of historical policy issues be resolved, including spectrum licensing, open access, universal service and data access. Ironically then, the solutions to these present problems will require that analysts consider revisiting pertinent historical issues in order to craft public policy that is meaningful for the providers, subscribers and technologies of today.

Note

1 Prior to the passage of the Telecommunications Act of 1996, the universal service fund was a mechanism for providing low-income households and high cost areas with telephone service via fees assessed to interstate long-distance carriers (FCC, 2015c). After the passage of the 96 Act, this definition of universal service was expanded to include health care providers in rural areas, eligible schools and libraries (FCC, 2015c).

References

12 Act. (1912). Radio Act of 1912. Public No. 264. Retrieved from http://earlyradiohistory. us/1912act.htm

34 Act. (1934). Communications Act of 1934. 47 U.S.C. § 151 et seq.

96 Act (1996). Telecommunications Act of 1996. Pub. LA. No. 104-104, 110 Stat. 56.

Abbate, J. (1999). *Inventing the Internet.* Cambridge: The MIT Press.

Abler, R. (1977). The telephone and the evolution of the American metropolitan system. In Pool, I.S. (Ed.), *The social impact of the telephone.* Cambridge: MIT Press.

Alleman, J., Rappoport, P., & Banerjee, A. (2010). Universal service: A new definition? *Telecommunications Policy, 34*(1), 86–91.

Allen, J.C., & Koffler, E.L. (1999). The Telecommunications Act of 1996: Its implementation in the US South. *Rural Development Issues Facing the South.* SRDC Mississippi State University.

Baller, J. (2014). State restrictions on community broadband services or other public communications initiatives. Baller Herbst Law Group. Retrieved from http://www.baller. com/wp-content/uploads/BallerHerbstStateBarriers1-1-14.pdf

Benjamin, L.M. (1982). *Deregulation? Early Radio Policy Reconsidered.* Paper presented at the Annual Meeting of the Association for Education in Journalism, 651st, Athens, OH, July 25–28, 1982. Retrieved from http://files.eric.ed.gov/fulltext/ED217431.pdf

Blackman, C.R. (1995). Universal service: Obligation or opportunity? *Telecommunications Policy, 19*(3), 171–176.

Brodkin, J. (2014a). Who wants competition? Big cable tries outlawing municipal broadband in Kansas. *Ars Technica.* January 31, 2014. Retrieved from http://tinyurl.com/ qjewxgy

Brodkin, J. (2014b). ISP lobby has already won limits on broadband in 20 states. *Ars Technica.* February 12, 2014. Retrieved from http://arstechnica.com/tech-policy/2014/02/ isp-lobby-has-already-won-limits-on-public-broadband-in-20-states/

Canon, S. (2014). Another Google Fiber delay in Kansas City, as the rest of the world catches up. *The Kansas City Star*, October 4, 2014. Retrieved from http://tinyurl.com/ o4ml564

Caristi, D. (2000). Policy initiatives and rural telecommunications. In Korsching, P.F., Hipple, P.C., & Abbott, E.A. (Eds), *Having all the right connections: Telecommunications and rural viability.* London: Praeger.

CBN. (2015). *Community Broadband Network. Community Network Map.* Retrieved from http://www.muninetworks.org/communitymap

CHRS. (2014). *100 Years of Radio.* California Historical Radio Society. Retrieved from http://www.californiahistoricalradio.com/radio-history/100years/

Clark, D.D., Gillett, S.E., Lehr, W., Sirbu, M.A., & Fountain, J.E. (2003). Local government stimulation of broadband: Effectiveness, e-government, and economic development. *E-Government, and Economic Development (January 2003). KSG Working Papers Series RWP03-002.*

Coase, R.H. (1959). The federal communications commission. *Journal of Law and Economics*, 1–40.

Compaine, B., & Weintraub, M. (1997). Universal access to online services: An examination of the issue. *Telecommunications Policy, 21*(1), 15–33.

CSUDH. (2014). *A short history of the telephone industry and regulation.* California State University Dominguez Hills. Retrieved from http://bpastudio.csudh.edu/fac/lpress/471/ hout/telecomHistory/

Crandall, R.W., Sidak, J.G., & Singer, H.J. (2002). The empirical case against asymmetric regulation of broadband Internet access. *Berkeley Technology Law Journal, 17*(3), 953–987.

Cukier, K.N. (1998). The global Internet: A primer, *TeleGeography* 1999. Washington, DC: Telegeography.

CyberTelecomm. (2013). Municipal broadband. *Cybertelecom Federal Internet Law & Policy An Educational Project.* Retrieved from http://www.cybertelecom.org/broadband/muni.htm

Dippon, C., Huther, C., & Troy, M. (2010). Replacement of the legacy high-cost universal support fund with a Connect America Fund. *Communications & Strategies, 80*, 67–81.

FCC [Federal Communications Commission]. (1997). FCC 1st Report and Order Universal Service Docket 96-45. May 8, 1997. Retrieved from http://tinyurl.com/mq3apo8

FCC [Federal Communications Commission]. (2010). *Connecting America: The National Broadband Plan.* Retrieved from http://download.broadband.gov/plan/national-broadband-plan.pdf

FCC [Federal Communications Commission]. (2015a). *Types of broadband connections.* Retrieved from http://www.fcc.gov/encyclopedia/types-broadband-connections

FCC [Federal Communications Commission]. (2015b). *Universal service program for schools and libraries (E-Rate).* Retrieved from http://www.fcc.gov/guides/universal-service-program-schools-and-libraries

FCC [Federal Communications Commission]. (2015c). *Universal service fund.* Retrieved from http://www.fcc.gov/encyclopedia/universal-service-fund

Ferguson, C.H. (2002). The U.S. broadband problem. Policy Brief #105-2002. Retrieved from http://tinyurl.com/k5o4vx2

Freese, D.D. (2002). Bad for broadband. *TCS Daily.* February 22, 2002. Retrieved from http://tinyurl.com/k5ttjnp

Gasman, L. (1998). Universal service: The new telecommunications entitlements and taxes. *Cato Policy Analysis, 310*, June 25, 1998. Retrieved from http://www.cato.org/pubs/pas/pa-310.html

Gillett, S.E. (2005). Municipal wireless broadband: Hype or harbinger. *Southern California Law Review, 79*, 561.

Gillett, S.E., Lehr, W.H., & Osorio, C. (2004). Local government broadband initiatives. *Telecommunications Policy, 28*(7), 537–558.

Gorman, S.P., & Malecki, E.J. (2000). The networks of the Internet: An analysis of provider networks in the USA. *Telecommunications Policy, 24*(2), 113–134.

Gross, G. (2015). States threaten lawsuit against Obama's municipal broadband plan. *PC World.* January 26, 2015. Retrieved from http://tinyurl.com/melkmg3

Grubesic, T.H. (2003). Inequities in the broadband revolution. *Annals of Regional Science, 37*, 263–289.

Grubesic, T.H. (2006). A spatial taxonomy of broadband regions in the United States. *Information Economics and Policy, 18*, 423–448.

Grubesic, T.H. (2012). The US national broadband map: Data limitations and implications. *Telecommunications Policy, 36*(2), 113–126.

Grubesic, T H. (2015). The broadband provision tensor. *Growth and Change. 46*(1): 58–80.

Grubesic, T.H., & Murray, A.T. (2004). Waiting for broadband: Local competition and the spatial distribution of advanced telecommunication services in the United States. *Growth and Change 35*(2), 139–165.

Hales, B., Gieseke, J., & Vargas-Chanes, D. (2000). Telecommunications and economic development: Chasing smokestacks with the Internet. In Korsching, P., Hipple, P.C, &

Abbott, E.A. (Eds.), *Having all the right connections: Telecommunications and rural viability* (pp. 257–275). Westport, CT: Praeger Publishers.

Hansell, S. (2005). Cable wins Internet-access ruling. *The New York Times*. June 28, 2005. Retrieved from http://tinyurl.com/mtlft9p

KCSV. (2015). *Kansas City Startup Village*. Retrieved from http://www.kcstartupvillage.org/

Kish, D. (2015). Google Fiber is coming to Atlanta, Charlotte, Nashville and Raleigh-Durham. January 27, 2015. Retrieved from http://tinyurl.com/kvv2uaa

Kolko, J. (2012). Broadband and local growth. *Journal of Urban Economics, 71*(1), 100–113.

Kruger, L.G. (2012). Background and issues for congressional oversight of ARRA broadband awards. Congressional Research Service. July 31, 2012.Retrieved from https://www.acuta.org/acuta/legreg/110714b.pdf

Kuschnick, B. (2001). The Bell monopolies are killing DSL, broadband, and competition. New Networks Institute. Retrieved from http://www.newnetworks.com/FinalCLECHarm.doc

Lanxon, N. (2008). The 50 most significant moments of Internet history. September 25, 2008. Retrieved from http://tinyurl.com/pqhmhay

Lasar, M. (2011a). How the *Titanic* disaster pushed Uncle Sam to "rule the air." *Ars Technica*. July 7, 2011. Retrieved from http://tinyurl.com/ne8fm22

Lasar, M. (2011b). The National Broadband Map: a $350 million "boondoggle"? *Ars Technica*. June 3, 2011. Retrieved from http://tinyurl.com/7pkae6f

Lehr, W.H., & Glassman, J.K. (2001). The economics of the Tauzin-Dingell Bill: Theory and evidence. *Center for e Business@ MIT-Paper, 128*.

Levin, S.L. (2010). Universal service and targeted support in a competitive telecommunications environment. *Telecommunications Policy, 34*(1), 92–97.

Mack, E.A. (2014). Broadband and knowledge intensive firm flusters: Essential link or auxiliary connection? *Papers in Regional Science, 93*(1), 3–29.

Matteson, S. (2013). Laying down the facts on Google Fiber. *Tech Republic*. April 26, 2013. Retrieved from http://tinyurl.com/nlqhq7u

MGF. (1935). Communications Act of 1934. *Virginia Law Review*, 318–325.

Miller, C.C. (2014). Why the U.S. has fallen behind in Internet speed and affordability. *The New York Times*. October 30, 2014. Retrieved from http://tinyurl.com/n6eb9nw

Moss, D.A., & Fein, M.R. (2003). Radio regulation revisited: Coase, the FCC, and the public interest. *Journal of Policy History, 15*(04), 389–416.

Mueller, M. (1993). Universal service in telephone history: A reconstruction. *Telecommunications Policy, 17*(5), 352–369.

Mueller, M. (1997). *Universal service: Competition, interconnection, and monopoly in the making of the American telephone system*. Cambridge: MIT Press.

National League of Cities (2013). *Number of municipal governments & population distribution*. Retrieved from http://tinyurl.com/kazojqy

National Research Council. (2002). *Broadband: Bringing home the bits*. Committee on Broadband Last Mile Technology, Computer Science and Telecommunications Board, Division on Engineering and Physical Sciences, National Research Council. Washington, DC: National Academies Press.

NCTA (2013). *The Rural Broadband Association*. Retrieved from http://tinyurl.com/kthookg

NTIA. (2015). *Broadband USA: Program information*. Retrieved from http://www2.ntia.doc.gov/information

Parker, E.B. (2000). Closing the digital divide in rural America. *Telecommunications Policy*, *24*(4), 281–290.

PBS. (2014). Radio in the 1930s. *History Detectives*. Retrieved from http://tinyurl.com/bu3g7nj

Phillips, H.F., & Tessler, J. (2003). FCC ruling is split decision for telecom rivals. *San Jose Mercury News*. Retrieved from www.siliconvalley.com/mld/siliconvalley/5228457.htm

Picot, A., & Wernick, C. (2007). The role of government in broadband access. *Telecommunications Policy*, *31*(10), 660–674.

Roosevelt Institute. (2010). Communications Act of 1934. *Next New Deal*. Retrieved from http://www.nextnewdeal.net/communications-act-1934

Rosston, G.L., &. Wimmer, B.S. (2000). The 'state' of universal service. *Information Economics and Policy*, *12*, 261–283.

S. 2430. (107th). *Broadband Regulatory Parity Act of 2002*. Retrieved from https://www.govtrack.us/congress/bills/107/s2430

Somer, J. (2015). What the net neutrality rules say. *New York Times*. Retrieved from http://tinyurl.com/kgaw2mg

Speta, J.B. (2000). Handicapping the race for the last mile: A critique of open access rules for broadband platforms. *Yale Journal on Regulation.*, *17*, 39.

Spiwak, L.J. (2015). *Internet law*. Retrieved from http://tinyurl.com/no4q7zg

Stricker, J. (2013). Casting a wider net: How and why state laws restricting municipal broadband networks must be modified. *The George Washington Law Review.*, *81*, 589.

Tauzin-Dingell Bill (H.R. 1542). H.R.1542 - Internet Freedom and Broadband Deployment Act of 2001. Retrieved from https://www.congress.gov/bill/107th-congress/house-bill/1542

Telephony Museum (2014). *Telephone history: The new century 1901–1940*. Retrieved from http://tinyurl.com/po4qotz

Thierer, A.D. (2001). The Tauzin-Dingell Bill and the National Academy of Sciences broadband study: Calls for broadband freedom. *Cato TechKnowledge* No. 29. Retrieved from http://tinyurl.com/loqqodg

Úrzua, C.A.O. (2004). *Bits of power: The involvement of municipal electric utilities in broadband services* (Doctoral dissertation). Massachusetts Institute of Technology.

Valdés, A. (1988). Explicit and implicit food subsidies: Distribution of cost. Baltimore, Maryland: John Hopkins University Press. Retrieved from http://tinyurl.com/l6k6oct

Varian, H.R., & Shapiro, C. (1999). *Information rules: A strategic guide to the network economy*. Cambridge: Harvard Business School Press.

Warf, B. (2003). Mergers and acquisitions in the telecommunications industry. *Growth and Change*, *34*(3), 321–344.

Willihnganz, J. (1994). Debating mass communication during the rise and fall of broadcasting. *BRIE Working Paper 74*. Retrieved from http://brie.berkeley.edu/publications/WP%2074.pdf

Wu, T. (2007) A brief history of American telecommunications regulation. *Oxford Encyclopedia of Legal History*. Available at SSRN: http://ssrn.com/abstract=965860

3   Broadband infrastructure

Readers may find it surprising that an entire chapter is dedicated to the topic of broadband infrastructure – the wires, tubes, software and associated hardware that makes integrated voice, data and video services possible. For most socio-economic and planning scientists and regional development experts, technological platforms and their inner workings are of little interest, particularly when compared to the more conspicuous policy issues surrounding broadband such as access equity, economic development, market competition and pricing. However, it is our view that ignoring the technological foundations of broadband and its associated infrastructure is a mistake for several reasons. First, issues of broadband access are strongly related to the type of platforms provisioned for a region. For example, without understanding the technological constraints of xDSL and its geographic service limitations (Grubesic, 2008), analysts can easily overestimate broadband penetration for a region (Grubesic, 2012a). Second, a similar argument can be made for broadband quality of service (QoS). Depending on the platform, QoS can quickly degrade with distance (Grubesic, Matisziw, & Murray, 2010) or be impacted by ingress noise and/or congestion on a neighborhood node for cable systems (Abe, 2000). Third, one of the most remarkable facets of broadband and its development over the past two decades is the relative dynamism of the access technologies available for end users. From digital subscriber lines, to hybrid fiber-coaxial (HFC) systems and Worldwide Interoperability for Microwave Access (WiMAX), access platforms continue to become faster, less expensive, more flexible and interoperable. A failure to acknowledge and better understand the importance of these platforms and their technical characteristics can lead to myopic policy development and suboptimal planning and management strategies.

The purpose of this chapter is to offer an overview of broadband access technologies, providing readers with a basic understanding of the most popular platforms, their strengths, weaknesses and technical limitations. This discussion is facilitated through the development of a basic typological framework for broadband infrastructure that uses both the operational and geographic characteristics of the access networks to deepen our understanding of the importance of broadband for regional development.

A hierarchy of access networks

The fundamental premise of telecommunication networks is to enable communications at a distance. Evolving from early efforts associated with the telegraph and the telephone (Fischer, 1992), the global telecommunication network of today is a mass of copper, fiber optic cables, hardware and software that is structured to operate in unison for achieving an exceedingly complex goal: connecting people and things to the network across vast stretches of geographic space.

Telecommunication network systems are organized in a hierarchical structure that reflects both operational and geographic goals and constraints for optimizing system performance, coverage, access and accessibility. Figure 3.1 provides a graphical summary of this hierarchy. At the top level, wide area networks (WANs) function as the core backbones for the telecommunication system, transporting both data and voice traffic between continents, countries and cities (O'Kelly & Grubesic, 2002), often using submarine links (Malecki & Wei, 2012). One example of a WAN is the Internet 2 network (Figure 3.2). It is an advanced hybrid optical and packet network designed and implemented to facilitate collaborative research on emerging network technologies. In addition to being a network technology test bed, Internet 2 also provides commercial peering services (Grubesic, Matisziw, & Ripley, 2011). The largest infrastructure providers in this group are known as Tier 1 carriers and include Level 3, Verizon, Sprint, AT&T

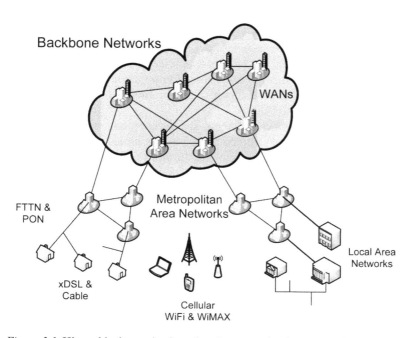

Figure 3.1 Hierarchical organization of a telecommunications network.

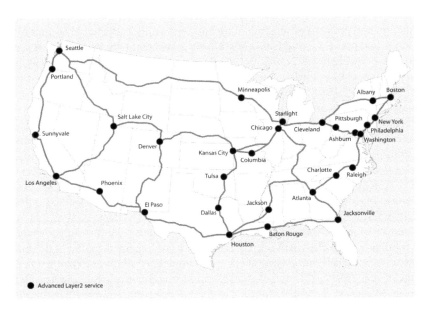

Figure 3.2 Internet 2 (Abilene) backbone network: Advanced Layer 2 services.

and TellaSonera amongst others. For reference, TellaSonera operates 43,000 km of fiber and 200 points of presence in 59 cities throughout North America, Europe and Asia (Peering DB, 2014). Unlike Tier 1 providers, Tier 2 carriers are more regionally oriented WANs that are often forced to purchase transit from Tier 1 providers to reach portions of the Internet. XO Communications is a good example of a Tier 2 provider in the U.S.[1]

Metropolitan area networks (MANs) are autonomous network systems composed of optical fibers that interconnect with the global backbone via network access points (NAPs). The geographic footprint of MANs is smaller than WANs, typically limited to metropolitan areas and perhaps a small portion of adjacent exurban areas. MANs are often classified as "middle mile" networks. Figure 3.3 displays a portion of the nearly 3,000 route miles of metropolitan network fiber operated by Integra Networks (Integra, 2014), focusing on the Portland, OR region.[2] MAN systems multiplex traffic from leased private lines, local area networks (LANs) and other access systems for transmission to the global backbone system.

Finally, LANs are made up of computers and other telecommunications equipment and devices from private organizations, public institutions, businesses or households that are sending and receiving traffic to MANs and the global backbone via access networks. These access networks serve as a bridge between end users and the global telecommunications system (MANs, WANs) and use leased private lines (e.g. OC-3 [about 50 Mbps] to OC-192 [about 10 Gbps]), twisted

Figure 3.3 Metropolitan area network: Integra network in/around Portland, Oregon.

copper pairs (e.g. DSL [digital subscriber line]), coaxial cable, hybrid coaxial or wireless systems for data collection and transmission. The geographic extent of LANs is smaller than MANs, typically covering 15 km or less. Again, tapping into these LANs are the "last mile" access platforms such as the DSL, cable, fiber and hybrid coaxial systems. These access technologies are the physical links between ISPs and end users, providing for a variety of applications and services such as integrated voice, data and video services to both residential and small- and medium-sized business segments.

Broadband is local

There are several reasons that local access technologies are so important to core issues of regional development. First, broadband is an inherently local/regional problem. Provision decisions are often tied to the demographic and socio-economic conditions of local markets (Gabel & Kwan, 2001; Grubesic & Murray, 2002; Oden & Strover, 2002; Horrigan, Stolp, & Wilson, 2006; Prieger & Hu, 2008; Mack & Grubesic, 2009). In short, providers are keen to rollout new technologies in markets where a fairly rapid return on investment can be realized. This means that the largest urban and suburban markets, with a greater density of affluent residents are usually the first locations to obtain new and/or advanced services. Conversely, smaller rural and remote markets are typically a generation or two behind the latest platform innovations because the density of consumers and the potential ROI is much lower (Grubesic, 2012a, 2012b). Second, providers have a long history of reusing and repurposing existing infrastructure for new applications (Moyer, 2009). This reduces investment costs, extends the useful life of capital investments and creates new sources of revenue for providers. Both DSL and cable are excellent examples of this. The retrofitting of these nearly ubiquitous local infrastructure systems to accommodate broadband has been a huge success for both the cable and telecommunications industries. Third, local markets are worth billions. Although hardwired residential connections are slowly losing favor with consumers as they continue to migrate to mobile platforms, the technological foundations of mobile markets remain inherently *local*. For example, mobile connections which use free-space optical networks, Wi-Fi mesh or WiMAX networks, cellular networks or satellite systems are eventually switched to the global backbone via local gateways. Simply put, the bulk of wireless systems remain hardwired, at least somewhere in their network architecture, but their wireless access points remain highly localized.

One final consideration pertaining to broadband is the concept of a first-last-mile bottleneck (Kazovsky, Cheng, Shaw, Gutierrez, & Wong, 2011). As detailed previously, although first-last-mile access technologies have greatly improved during the past decade, they remain the proverbial "weak link" for the global telecommunications network. In particular, platforms that rely on copper (e.g., xDSL and cable) are quickly becoming outmoded. There is a more complete description of xDSL and cable platforms later in this chapter, but consider the most high bandwidth version of digital subscriber line technology, VDSL2, which provides downstream speeds of 100 Mbps and upstream speeds of 30 Mbps (Kazovsky et al., 2011). When compared to the most basic broadband passive optical network (BPON), which offers speeds of 622 Mbps downstream and 155 Mbps upstream, copper platforms such as xDSL (in any variety) simply cannot compete. Further, even when BPON fiber is divided using an optical splitter to serve multiple users, the available bandwidth in fiber systems still exceeds VDSL2 for end users.

The reason that the first-last-mile bottleneck is such a concern to both consumers and providers is that residential subscribers and small businesses are demanding more bandwidth and better QoS for real-time video applications, distance learning, gaming and big-data analytics. More importantly, the world is shifting away from an internet of computers, toward an internet of "things" (Mattern & Floerkemeier, 2010). With internet-enabled refrigerators, thermostats and garage door openers in homes, combined with the supervisory control and data acquisition systems (SCADA) used to remotely operate and monitor much of the critical infrastructure in the United States (Alcaraz, Fernandez, & Carvajal, 2010; Murray & Grubesic, 2013), the bandwidth required to connect these "things" is going to increase exponentially over the next decade. Existing copper-based broadband platforms, such as xDSL and cable will almost certainly be insufficient for this increased demand, so alternative platforms such as fiber will gain favor. In the next section, we profile the complex mix of broadband platforms, detailing their strengths, weaknesses and quirks, along with providing a bit of educated speculation about their long-term viability in an increasingly competitive broadband landscape.

Digital subscriber lines (xDSL)

Digital subscriber lines are a copper-based broadband platform that allow for the overlay of a high-capacity data channel on top of the standard analog voice channel on regular telephone lines. Digital subscriber lines maintain a 34% share of the fixed broadband market in the United States (IHS, 2013), with approximately 31 million connections. The xDSL ecosystem is somewhat complex, relying upon end-user modems, the copper transmission lines and switching equipment (digital subscriber line access multiplexers [DSLAM]) located within the local central office (CO). In some cases, DSLAMs are deployed remotely to extend the reach of xDSL to locations outside typical service ranges (Grubesic & Horner, 2006). Figure 3.4 illustrates a typical DSL service architecture, which is enabled by a modem at the end user's location (e.g., household). The modem is connected to a long run of copper wire that ends at the local DSLAM and/or CO. Once the data reach the CO, they are routed to a MAN or WAN for transmission to the global backbone.

There are a number of different xDSL access technologies. Asymmetric DSL, or ADSL, technologies are optimized for both residential and small business use. In addition to supporting plain old telephone service (POTS), ADSL can simultaneously support digital data transmission. The basic premise for asymmetry in this technology is that end users typically require more downstream than upstream bandwidth. As a result, typical ADSL connections average downstream speeds of 8 Mbps with upstream speeds averaging 1–2 Mbps. Second generation technologies, such as ADSL2, offer more bandwidth for end users (~12 Mbps and 4 Mbps, respectively) due to improvements in modulation techniques. Most providers try to limit loop lengths (i.e., distance from end-user premises to the CO) to about 3.6 km (12,000 ft.) to minimize attenuation and ensure service quality. These infrastructure challenges will be discussed later in this chapter.

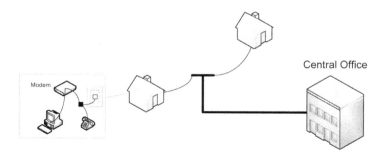

Figure 3.4 Typical xDSL service architecture.

Very-high-bit-rate DSL (VDSL) is a related technology that provides more bandwidth over copper lines and uses a different frequency band than ADSL. VDSL supports high definition television, internet and voice over IP services via a single connection. However, the geographic range of VDSL is quite limited, extending only 1.5 km (~5,000 ft.) from a CO. VDSL2 is a second generation technology that uses frequencies up to 30 MHz to provide symmetric bandwidth of 100 Mbps. The service range for VDSL2 is extremely limited, extending only 300 m (~1,000 ft.) from a CO. Both versions of VDSL make use of remote DSLAMs with some regularity.

A final xDSL technology worth noting is SHDSL, or symmetrical high-speed digital subscriber line. Again, leveraging improvements in modulation technology and sharing frequency with POTS, SHDSL can transmit and receive data at 5.696 Mbps. However, because it uses the frequency that POTS is transmitted on, SHDSL cannot support traditional telephone service on the same line.

As detailed previously, xDSL represents a repurposing of existing local infrastructure for digital data transmission. Not surprisingly, such efforts are fraught with challenges and several significant limitations for xDSL broadband service provision have emerged. On the positive side, the global popularity of xDSL is unrivaled. Of the 321 million broadband subscribers in OECD countries, 52.7% are connected with digital subscriber lines (OECD, 2013). Second, xDSL offers excellent physical security. Each telephone line connects to only one customer and the bandwidth offered by xDSL platforms is dedicated. When compared to cable broadband, which is a shared access system, this is a decided advantage. As detailed by Gorshe, Raghavan, Starr, & Galli (2014), shared systems are susceptible to a small set of users consuming more than their share of the bandwidth. This is less common with xDSL, however, traffic from multiple DSL lines is multiplexed together within a DSLAM. Thus, if the DSLAM or its associated uplink is under-capacitated, there may be problems, but this is rare.

Although the benefits of point-to-point broadband service are a distinct advantage for xDSL technologies, there are several notable limitations as well. First, standard telephone lines are constructed of two single core copper wires that are twisted around each other. As detailed by Grubesic and Murray (2002), the bulk of this infrastructure, particularly in residential areas, dates back to the 1950s. In many instances, this requires xDSL providers to "condition" lines prior to service. The conditioning process is required for one (or more) reasons. First, load coils must be removed to provide high-quality xDSL service. Load coils are passive (unpowered) devices that lengthen the distance voice can travel over twisted copper pairs. They are usually placed on the local loop at intervals of 3,000–6,000 ft (~900–1,800 m) to increase the fidelity of a voice signal. Without them, a voice signal can become crippled by its own reflection in the copper wires (Abe, 2000; Cartwright, 2000). However, because load coils are suppressing this noise at high bandwidth, the transmission of xDSL signals becomes difficult. Bridge taps also need to be removed. Bridge taps are "accidental" connections of an additional local loop to the primary local loop. There are millions of bridge taps located in residential areas and business parks throughout the U.S. When a phone company runs copper cable down a residential street, the length of the cable serving any given household often extends well beyond the residence it serves. In other words, the installer has more cable than needed. Instead of cutting that cable, installation technicians often leave the extra length in case they have to use it later for another household. Installers are then able to take the wires from the household in question and tap (via splice boxes) the main copper cable serving a residential street, subsequently running back to the CO (Abe, 2000). The extra wiring left in place during installation can severely impede xDSL transmission quality to a household, often acting as an antenna and picking up all kinds of electronic noise from the ambient environment (ibid.).

Finally, the distance constraints associated with xDSL service range are particularly notable when exploring issues of regional development, broadband access, equity and QoS. As detailed previously, the availability of the most far-reaching technology in the xDSL family, ADSL, is generally limited to about 3.6 km from a CO. In part, this is a structured geographic service restriction initiated by providers. Simply put, as line length increases, the quality of xDSL service decreases and most providers are unwilling to compromise QoS to add to their subscriber base. Some unhappy subscribers would certainly argue with this assertion, but in aggregate, these limitations have been an asset to providers throughout the U.S., even with the xDSL market in decline. On the whole, xDSL delivers promised speeds (FCC, 2013), although there is certainly some heterogeneity amongst markets (Grubesic, 2015). The core policy problems related to these service limitations are detailed in Grubesic (2008, 2012a) and Grubesic et al. (2011). First, xDSL access is chronically overestimated because most policy analysts do not have access to wire-center service area data, central office location data or the appropriate tools to estimate local loop lengths from central offices or remote DSLAMs to accurately estimate service reach. Second, variability in QoS associated with xDSL is a major policy and provision issue. Subscribers located more closely to central offices and their associated DSLAMs often experience higher broadband service quality. Those located farther away can experience poorer QoS and diminished download speeds (Grubesic et al., 2011). In this context, the quality of one's experience "online," both in terms of speed and stability, dictates the types of activities in which the end user can engage. Internet applications with higher bandwidth demands (e.g., streaming video) can be limited for xDSL subscribers located on the periphery of their wire-center service area. Worse, it is likely that residential subscribers are paying identical subscription fees for xDSL service, regardless of location or performance, unless some type of differentiated service program is in place (ibid.).

In sum, xDSL is a popular and widely provisioned access platform that is unlikely to disappear anytime soon. However, with market share shrinking, at least in the United States (FCC, 2013), and the limited reach of its most affordable platforms, xDSL's best years may be in the past. There are glimmers of hope, however. Incumbent local exchange carriers (ILECs) are sensitive to these changes in the market and improvements to xDSL technology are helping extend its lifespan as an access technology. For example, Gorshe et al. (2014) suggest that repeater technology is helping mitigate the geographic service restrictions for some consumers. Specifically, if a repeater is placed in the middle of a long copper line, it is able to amplify the signal and transform the line, at least from an operational perspective, into two "shorter" segments. Thus, the geographic reach of xDSL is extended. Bandwidth is also being increased using bonding technologies that connect several telephone lines with multiple xDSL modems. Data are then sent across multiple lines, simultaneously, enhancing bandwidth capabilities and transforming xDSL into a "multi-lane highway" (Gorshe et al., 2014). So, while xDSL is certainly not dead, and there are glimmers of hope for this copper-based technology, it is unclear if xDSL can continue to accommodate end users increasing demand for bandwidth as alternative technologies such as fiber become more ubiquitous.

Hybrid fiber-coaxial cable (HFC)

The dominant broadband access platform in the United States is HFC. HFC maintains more than a 50% share of the fixed broadband market in the United States (IHS, 2013), with providers adding an average of 600,000 connections each quarter since 2011 (ibid.). One reason that HFC has such strong momentum in the U.S. is the relative ubiquity as a network. This is not to say that cable systems exceed the reach of the POTS network, but, unlike xDSL systems, there are no severe distance constraints with HFC. Cable systems were designed, structured and deployed as networks to optimize broadcast television traffic to both urban and rural/remote areas. A second reason, strongly rooted in broadband economics, is the ability for cable providers to offer "triple play" packages, where voice, video and data connections are bundled for consumers (Green, 2002; Schilke & Wirtz, 2012). This synergy of offerings from cable providers has the potential to create new revenue streams, including media sales, advanced home networking, unified software and applications as well as remote service management (Ulm & Weeks, 2007). As detailed previously, with a market share of over 50%, consumers in the United States have clearly embraced both cable and bundled service packages, even if these packages cost significantly more in the U.S. than elsewhere (Cassidy, 2014).

During the early broadband deployment, the major challenge for cable operators was converting their systems to handle bidirectional traffic. Specifically, the network repeaters that were prevalent throughout cable systems in the United States were structured to handle downstream traffic only.[3] Once this conversion process was complete, cable networks were able to handle both downstream and upstream traffic. Figure 3.5 illustrates a typical HFC network architecture. End users maintain a data over cable service interface specification (DOCSIS) modem on their premises. This modem is connected to a run of cable that multiple houses in a neighborhood share.[4] For more remote premises, the cable passes through an amplifier on its way to a local fiber node. Once the signal reaches the fiber node, its path across copper is complete and it is routed to a distribution hub and eventually a primary hub that connects to a master headend. At this junction, all traffic is routed to large MANs and/or WANs. The master headend also maintains

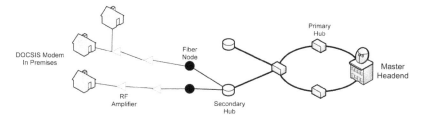

Figure 3.5 Typical HFC network architecture.

connections to the public switched telephone network and satellite feeds for television content distribution. HFC networks are strongly hierarchical in structure and these types of agglomeration networks provide for an efficient platform for content distribution and consumption.

HFC systems rely on the DOCSIS protocol for managing both physical and medium access control (MAC) layers. As noted previously, HFC systems are shared, with many subscribers using the local network. DOCSIS helps specify how both downstream and upstream data is carried through the network, including managing allocation to radio frequency channel slots, line coding and modulation (Abe, 2000; Gorshe et al., 2014). More importantly, DOCSIS also specifies the appropriate encryption protocol to ensure that an end user cannot access data from other end users within this shared system. It also uses authentication protocols to ensure that end users are able to access only the services to which they are entitled (Kazovsky et al., 2011).

The advantages of HFC as a broadband access technology are numerous. First, as noted earlier, the ability to offer bundled services (e.g. voice, video and data) has proven to be a winning combination for multiple system operators (MSOs) in the United States (Green, 2004), attracting hundreds of thousands of subscribers each quarter. Second, when compared to the twisted pair infrastructure in POTS, coaxial cables have a much higher bandwidth (1,000 MHz), enabling them to support a much higher data transfer rate (Kazovsky et al., 2011). For example, if the signal-to-noise ratio exceeds a value of 30 on a coaxial cable, then typical downstream transmission speeds can approach 40 Mbps and upstream data transfer rates can approach 10 Mbps, far exceeding standard ADSL bandwidth. In other instances, high-performance broadband packages allow for downstream transmission speeds that approach 100 Mbps. Third, DOCSIS supports service up to 160 km between distribution hubs or headends and the cable modem at a business or residential premise (Gorshe et al., 2014). Although this distance represents an extreme case, and distances of 24 km are more typical, this is a decided advantage over xDSL technologies.

HFC is not without limitations. As noted previously, HFC systems are shared networks and the available bandwidth for a service neighborhood is a community resource. Therefore, power-users or "bandwidth hogs" can often use more than their share of bandwidth on a cable system if they remain unchecked. Not surprisingly, there is a strong temporal component to this limitation. At 3am, traffic on the cable network is low and advertised bandwidth speeds would be available to any end user on the system. However, during peak use times (~ 6pm – 10pm) advertised speeds may not be available because of the heavy use of the system by local subscribers. A second limitation is that there remain some locations in the United States that do not have HFC systems available. As detailed previously, the geographic reach of cable is significant, but not all providers have updated their systems to handle bidirectional transmissions. As detailed in Chapter 5, there are gaps in provision for HFC in many locations throughout the United States. Finally, Gorshe et al. (2014) note that the legacy of video spectrum assignments in cable systems severely limits the bandwidth available for upstream transmissions from end users. As demand for upstream performance continues to grow, this will become a serious problem for HFC systems and their operators.

In sum, HFC is the dominant fixed terrestrial broadband access platform in the United States. Although its costs are relatively high for subscribers, end users are generally satisfied with cable broadband service (JD Power, 2013). Although the future is promising for HFC systems, there are some clouds on the horizon. With increasing competition from providers offering fiber-to-the-node (FTTN), cable broadband systems will need to rely on technological improvements such as channel bonding to match the downstream performance of FTTN deployments.[5] However, channel bonding may not be the only and/or best strategy for MSOs. Again, efforts to increase upstream transmission performance and the exceedingly high downstream bandwidth of FTTN may require HFC providers to explore all-fiber solutions, including FTTN to enhance first-last-mile bandwidth.

Fiber to the curb/home/building/node (FTTx)

FTTN systems represent one of the most promising developments in first-last-mile technologies for broadband providers for quite some time. As detailed previously, fiber connections offer the ability to eliminate bandwidth bottlenecks, enhance QoS and provider end users with a viable alternative to xDSL and HFC broadband platforms.

One family of FTTx technology is passive optical networks (PONs). PONs are fiber systems that do not maintain any active elements (i.e., powered electronics) along the fiber itself. Instead, active devices are located in the local CO or at the end-user premises. The major advantage of using passive systems is the ability to create a bidirectional point-to-multipoint network. As detailed by Gorshe et al. (2014), it is cost prohibitive to provide each end user with a dedicated fiber. Thus, the use of beam splitters for PONs allows providers to divide a single feeder fiber into 16, 32 or potentially 64 distribution fibers for some systems (Kazovsky et al., 2011). Each distribution fiber can then serve an individual household or business, allowing providers to create economies of scale and generate a higher return on investment. Additional savings are realized in this process because providers are able to share equipment amongst subscribers – a single expensive laser is used at the CO, while less expensive and lower quality lasers are allocated to each end user. It is also important to note that PON systems transmit both upstream and downstream data along the same fiber.

Figure 3.6 illustrates the typical network architecture for PONs, but it is important to note that several different types of PONs are currently deployed in the United States and elsewhere, each with their own standards and associated quirks. For example, BPONs are based on asynchronous transfer mode (ATM) signaling and transport protocols.[6] It also supports scalable traffic management and strong quality of service controls (Kazovsky et al., 2011). BPON offers data transfer rates of 155 or 622 Mbps to end users, but is geographically limited to a service distance of 20 km or less from a CO. Gigabit-capable PONs (GPON) use a combination of IP-based protocol and ATM or GPON encapsulation method (GEM) encoding and transporting data. Virtually all of the FTTN expansions currently underway in the United States use GPON (Gorshe et al., 2014). The advantage of GPON systems is their high data throughput, accommodating rates of 2.5 Gbps. GPON is also backwardly compatible with BPON. GPON service is geographically limited to 60 km or less from a CO. Finally, Ethernet PONs (EPON), which are widely deployed in portions of Asia (notably Japan) but are also making inroads to the United States, provide an effective data rate of 1 Gbps (with a maximum of 1.25) and use packet-based transmissions for all network activity (voice, video and data). Although EPON only allows for feeder fibers to be split into 16 distribution fibers, the use of Ethernet-based hardware allows EPON to maintain a slight cost advantage over GPON (Kazovsky et al., 2011) because no ATM equipment is required. EPON service is geographically limited to 20 km or less from a CO.

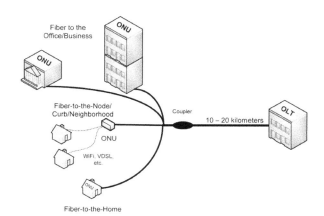

ONU: Optical Network Unit
OLT: Optical Line Terminal

Figure 3.6 Typical PON network architecture.

The key advantages to the use of FTTx are well known, but are particularly compelling for passive optical systems. The lifetime costs of all-fiber solutions are significantly lower than copper-based networks. Copper systems require expensive, finite lifetime electronic equipment with redundant power supplies located throughout a service region (Green, 2002). As detailed earlier, other than splitters, PONs require no equipment along feeder lines and, because of their passive nature, no powered electronics are required anywhere other than the CO and end-user premise. Second, in many ways, fiber is "future-proof." As broadband transmission technologies continue to improve, there is no reason that fiber needs to be replaced for a service region. Instead, these advances can be directly integrated into the CO or end-user equipment. Also, unlike xDSL or cable broadband systems, the low attenuation of fiber allows for a greater geographic reach of service, often extending 20 km or more from a central office (Gorshe et al., 2014). Further, all-fiber systems never display the patchwork infrastructure characteristics that many alternatives exhibit. There are no antiquated lines, amplifiers or cable types that increase the complexity of installations. Finally, fiber is more robust to extreme events than copper. Historically, companies such as Verizon and other major telecommunications operators relied on pressurized cables to keep water out of the system, often pumping nitrogen through the cable to prevent moisture from reaching the copper. However, during the flooding caused by Hurricane Sandy in 2012, 95% of Verizon's copper system was damaged, even with such safety measures in place. Ultimately, Verizon's response was spending billions of dollars replacing copper with fiber after the storm because it provides a more reliable infrastructure platform (Bonomo, 2012).

There are a few disadvantages associated with fiber worth noting. First, although overall costs continue to drop, the installation of new fiber is a significant capital expense for providers. Obviously, the return on investment can be quite high, but it is important to mention that capital costs are involved. Second, there are costs associated with upgrading test equipment for PONs. This includes the purchase of optical time domain reflectometers, which cost about $10,000 per unit.[7] Third, although fiber is more robust to water damage, it is still susceptible to physical damage. A running joke in the ISP community is that nefarious "fiber seeking backhoes" are constantly trying to cut or dig up critical backbone links. Finally, there are minor ecological threats to fiber systems. Birds often find that the protective Kevlar sleeves used for fiber systems are excellent nesting materials, rodents have gnawed their way through fiber (Bloomberg, 2001) and some plants wrap their roots around cable tightly enough to disrupt transmission capabilities (Goleniewski, 2002).

Overall, fiber is the most solid option for terrestrial broadband in the United States and elsewhere for the near future. Again, although the initial installation costs are of some concern, growing demand for bidirectional bandwidth and increased QoS will precipitate a larger provider base, as well as more end users. The one caveat with an all-fiber future is that providers will be hesitant to install FTTx networks in locations where the potential return on investment is low. This means that exurban and rural areas may need to rely on alternative broadband access technologies. In the next section, we explore some of these alternatives, emphasizing wireless options.

Wireless broadband access technologies

There are scores of wireless broadband platforms that are currently in use or being developed for commercial rollout. Gorshe et al. (2014) use a simple typology to categorize wireless technologies, *short-range* and *long-range*. Short-range wireless platforms, such as Wi-Fi, primarily evolved to provide access to data networking services in localized areas. Long-range platforms, such as cellular services and WiMAX, are structured to provide connectivity over widely distributed geographic areas. A second point of differentiation is between *fixed* and *mobile* technologies. Fixed wireless technologies include Wi-Fi and WiMAX (Vaughan-Nichols, 2004; Abichar, Peng, & Chang, 2006) use unlicensed spectrum and usually require a direct line-of-sight between the wireless transmitter and receiver to ensure service. Failing a direct line-of-sight, end users must be in close proximity to make a connection. Mobile wireless connections provide broadband access to mobile objects, such as pedestrians, automobiles, boats and trucks. Mobile connections also use licensed spectrum that is dedicated to a specific provider and include platforms such as 3GPP Long Term Evolution (LTE) and CDMA2000 (EVDO) amongst others (Agashe, Rezaiifar, & Bender, 2004; Dahlman, Parkvall, & Skold, 2013).

Wi-Fi

Where the short-range technologies are concerned, consider Wi-Fi, which is based on the IEEE 802.11 standards, and provides wireless connectivity for multiple devices in a household, small business or public space. Wi-Fi uses unlicensed spectrum and has restrictions on transmission power because it is designed to coexist with other wireless technologies in operation (e.g., wireless phones, cellular systems and radio-frequency identification systems). In standard implementations, a single Wi-Fi access point is connected to a terrestrial broadband platform (e.g., cable) and allows for point-to-multipoint wireless communications in a localized area. For more expansive implementations, Wi-Fi mesh networks use multiple interconnected access points to create a network, enabling automatic topology discovery, dynamic routing and a larger geographic footprint of service (Lee & Murray, 2010; Shillington & Tong, 2011). Wi-Fi has gained such a foothold in the current broadband landscape because it functions with a simple architecture, the equipment is inexpensive and Wi-Fi is relatively easy to configure and install. Similar to other wireless technologies, there are a variety of Wi-Fi standards that include *a*, *b*, *g*, *n* and *y*, all with unique coverage abilities and associated constraints (Table 3.1). Kazovsky et al. (2011) note that Wi-Fi is somewhat hampered as a first-last-mile technology because mesh networks that support multi-hop paths will be necessary. Also, radio frequency interference can limit bit rates for multi-hop systems. This is exactly why 802.11y technology is currently under development. As detailed in Table 3.1, the *y* standard will allow for data transfer rates of 54 Mbps, multi-hop paths and a coverage distance of nearly 5 km outdoors.

The limitations associated with Wi-Fi are worth detailing. First, Wi-Fi data transfer speeds are generally reliable if users are within range of the access point. Physical proximity increases the signal-to-noise ratio and helps devices connect and perform without problems. However, 50 Mbps rates for a Wi-Fi connection are constrained by the actual capacity of a broadband connection. For example, if end users have a Wi-Fi access point connected to a 10 Mbps ADSL service, the 50 Mbps transfer rate of the Wi-Fi hub is meaningless. Connection speed is constrained by the ADSL service, not the Wi-Fi hub. It is also important to note that Wi-Fi is adversely impacted by concrete, metal, large buildings and other obstacles that create a physical interference of the radio signal (Agrawal & Zeng, 2015).

WiMAX

Long-range technologies are quite varied and include the 802.16 standard, WiMAX. Similar to Wi-Fi, WiMAX also provides a low-cost solution for broadband access, but requires a larger base station for operation. WiMAX was initially designed to provide long-range fixed wireless for entire metropolitan areas, but its application has now morphed into more localized deployments in exurban and rural communities where installing terrestrial systems, such as fiber or xDSL, is viewed with skepticism by providers because of potentially low returns on investment. As a point-to-multipoint platform, WiMAX can support data transfer rates up to 75 Mbps within 50 km of a base station. However, much like cable, this bandwidth is a shared resource. As a result, most WiMAX systems support a realized bandwidth of approximately 1 Mbps to residential consumers and 2–3 Mbps for business users (Kazovsky et al., 2011). A new standard, 802.16m, purportedly supports rates up to 1 Gbps, but the transmission range may be less than 50 km (Nafea, Zaki, & Moustafa, 2013). WiMAX networks can also support a mesh topology, allowing for greater geographic coverage and faster deployment, but this comes at the expense of reduced data rates to end users (Kazovsky et al., 2011). In general, because WiMAX uses time-division multiplexing (similar to

Table 3.1 IEEE 802.11 Standards

Parameter	802.11a	802.11b	802.11c	802.11n	802.11y
Operating frequency, GHz	5	2.4	204	2.4 & 5	3.7
Maximum data rate (Mbps)	54	11	54	248	54
Maximum indoor trasmission distance (meters)*	35	40	40	70	50
Maximum outdoor transmission distance (meters)*	100	120	120	250	5000

what cellular carriers used prior to GSM), it is more efficient at allocating band-width than Wi-Fi systems (ibid.).

It is also worth noting that local multipoint distribution service (LMDS) and multichannel multipoint distribution service (MMDS) are related to WiMAX and were developed by the IEEE 802.16.1 working group. However, WiMAX out-performs both LMDS and MMDS, is more cost effective and has more market potential. As a result, it is likely that MMDS and LMDS will become obsolete (Kazovsky et al., 2011).

Cellular

Cellular systems also represent a popular long-range technology for wireless broadband access. Initially designed for providing mobile voice access, the first generation cellular systems were not structured to handle data. Today, both third and fourth generation (3G and 4G) systems handle more data than voice traffic (Higginbotham, 2010). In fact, recent estimates from Cisco suggest that wire-less networks in North America carried 740 petabytes of data per month during 2014, a 40-fold increase from 2009, where the system handled approximately 19 petabytes per month (Brown, 2011). This growth in data traffic can be attributed to improvements in network data handling capabilities. As 3G and 4G systems became more ubiquitous, demand for data services has increased. This demand, however, is not strictly driven by smartphones. The advent of LTE capable lap-tops and tablets is also fueling the use of cellular data networks.

The core architecture of cellular systems (Figure 3.7) includes base transceiver stations (BTS) and their associated cells, base station controllers (BSC), home location registers (HLR), visitor location registers (VLR) and mobile switch-ing centers (MSC) which connect directly to MANs and WANs via edge routers (Kazovsky et al., 2011). Cells are the base geographic unit of a cellular system and obtained their name from the typical honeycomb pattern of cell site installa-tions in places where cellular coverage is available (Newton, 2002). The size of cells varies based on the transmission power of the radio transmitters and receiv-ers. Macro-cells often have a radius up to 30 km, with micro- and pico-cells covering up to 2,000 m and 200 m, respectively. BTSs are located in the center of cells and host all of the equipment required for transmitting and receiving radio signals. BSCs are designed to allocate radio channels to end users for multiple BTSs. BSCs also function as a switch on the cellular network for concentrating many low-traffic density connections upstream to MSCs. MSCs are responsi-ble for routing voice calls to the public switched telephone network, as well as dealing with hand-off requirements between BSCs during a call, although this is sometimes handled by the BSCs in some carrier's networks.[8] MSCs are also responsible for data services between the cellular system and edge routers for the global internet backbone. HLRs are designed to obtain data for each mobile device requesting access to the network and to track the subscriber identity mod-ule (SIM) and associated telephone number for each mobile device.[9] The VLR is similar to an HLR, but is used for mobile devices and/or cellular subscribers that

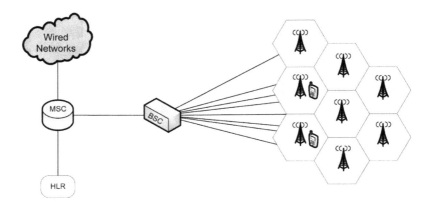

MSC: Mobile Switching Center

BSC: Base Station Controller

HLR: Home Location Register

Figure 3.7 Typical cellular system architecture.

are roaming. VLRs authenticate use of the roaming network using a registration notification system.

Data transfer rates for cellular systems vary considerably, at least in the United States. In a recent study, RootMetrics tested the 4G network performance of three major U.S. carriers (AT&T, Verizon and Sprint; Graziano, 2013). Results suggested that AT&T had the fastest network, with average download speeds of 14.3 Mbps and upload speeds of 8.5 Mbps. Maximum download speeds for AT&T approached nearly 50 Mbps. Verizon was a close second in terms of network speeds, but maintained coverage in all 77 markets tested, whereas AT&T only maintained a presence in 47 of the markets. Sprint placed third amongst the carriers, with average LTE download speeds of 10.3 Mbps and upload speeds of 4.4 Mbps. More importantly, Sprint maintained service in only 5 of the 77 markets tested. T-Mobile, a fourth major carrier in the United States, is now rolling out their 4G LTE network, but the active HSPA+ for T-Mobile provided 7.3 Mbps average downloads and uploads of 1.5 Mbps.

Aside from data caps and the cost of mobile broadband data plans, the obvious limitation with cellular-based broadband access is geographic coverage.[10] Although carriers such as Verizon and AT&T routinely tout the geographic expanse of their network with comprehensive maps (http://www.verizonwireless.com/wcms/consumer/4g-lte.html), the reality is that these maps represent a stylized, theoretical representation of coverage according to software programs that estimate spectrum propagation. Without a comprehensive and detailed testing system that is monitored by a third party for accuracy, providers can basically put anything they want into a map and not be accountable for any errors or misleading information. To combat this, an entire mini-industry has recently developed that is dedicated to crowd-sourcing poor cellular network performance. For example, deadcellzones.com uses consumer feedback to identify cellular phone reception problems. OpenSignal.com provides consumers with a free Android or IOS application that helps track cell signal strength data and then compiles this information for a more accurate representation of provider coverage. As of May 2014, over 825 cellular networks, 824,000 towers and 5.1 billion signal readings had been processed. In short, cellular coverage is an issue and will remain as a constraint for cellular broadband access into the conceivable future.

Satellite broadband

A final wireless access platform worth detailing is satellite broadband. Satellite broadband access technologies maintain a relatively small market share in the United States, with approximately 1.4 million subscribers in 2013 (Tauri Group, 2013). Providers such as Hughes, Wild Blue, Spacenet and ViaSat offer competitively priced packages for rural and remote subscribers in the U.S. For example, ViaSat's Exede internet service offers end users single rate, tiered service plans with data caps at 10 GB, 15 GB or 25 GB per month. Advertised download speeds are 12 Mbps downstream and 3 Mbps upstream. Packages also include unlimited downloads between 12am and 5am local time (ViaSat, 2014).

The architecture of satellite broadband systems is relatively simple, but the required equipment is more significant than a typical terrestrial broadband platform. End users require a satellite modem, which handles data encoding and modulation. They also require a satellite transceiver, which includes frequency converters and amplifiers, and an antenna, which is identical to a satellite television antenna used widely in the U.S. From an operational perspective, a geostationary orbital satellite is used to provide coverage to end users in an extremely large geographical area, often covering approximately 40% of the Earth's surface. This satellite functions as a repeater and establishes a wireless link between two points on the Earth's surface, the end user and a central office/headend. The satellite receives signals from stations on the Earth's surface as an *uplink*, amplifies the signals and transmits them back to Earth with a different frequency as a *downlink*. As one might expect, the major caveat with satellite service is significant signal attenuation. But, if latency is not a significant concern for end users, these delays are not crippling. There are also issues with rain fade, where atmospheric moisture (e.g., snow or rain) disrupt and/or degrade the connections between end users and the satellite (Hogg & Chu, 1975).

The promise of satellite broadband, at least for rural and remote areas, is that a relatively fast connection can be available in places where terrestrial providers or cellular companies have no plans for expansion. The truly remote portions of the United States and Canada are good examples in this context. It is simply too expensive to provide fiber, cable, xDSL or cellular service to customers at the bottom of a steep river canyon or those located hundreds of miles away from any major urban centers (Grubesic, 2012a). The problem with satellite broadband is that its coverage is far from ubiquitous. Further, even for locations that receive satellite coverage, there are significant variations in quality. In many ways, these operational constraints are quite similar to those detailed with xDSL and cable (e.g., geographic constraints and QoS issues), albeit for a wireless system. To clarify this point, consider Figure 3.8 that highlights the ViaSat beam classification map. There are two different coverages available, Exede 5 and Exede 12. Several areas are covered by both. As detailed previously, ViaSat offers 12 Mbps service in the Exede 12 areas and only 5 Mbps in Exede 5 regions. This is a big difference in performance and, more importantly, there are notable gaps in coverage throughout many of the most remote portions of the U.S., including Wyoming, portions of the Dakotas, Idaho and Montana. The point being made here is that satellite broadband, while effective for some remote areas is not a panacea – issues of availability and QoS remain.

Conclusion

There are other broadband access technologies that have not been addressed in this chapter. For example, in the early 2000s, there was some excitement concerning the prospects of power line communications for delivering broadband services. Unfortunately, power line communications suffer from significant operational challenges and the development of this platform has been quite slow relative to cable, fiber and xDSL (Gorshe et al., 2014). In short, companies simply could not deliver the bandwidth required to make it a competitive alternative for end users (Courtney, 2013).

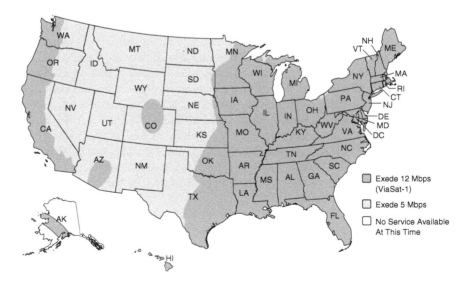

Figure 3.8 ViaSat broadband availability, 2014.

Rather than detail a host of fledgling technologies, this chapter focused on the core platforms that are currently available in residential and business markets in the United States and beyond. However, as detailed at the beginning of this chapter, most regional development experts, socio-economic planning scientists and policy analysts remain largely ambivalent about the technological facets of broadband access platforms and their provision challenges. In much of the published work on broadband, analysts only care about whether broadband is available or not. This is certainly an important question because it allows for additional work concerning equity, access, competition and larger policy concerns about the digital divide, e-commerce and participation in social media feasible. Further, the technological complexities associated with broadband access platforms make their analysis more challenging. Researchers without a background in engineering or experience in the telecommunications industry often find this technical material impenetrable and difficult to interweave with more traditional econometric or policy-based analyses.

This disconnect is unfortunate because it leads to misinformed analytical efforts that are largely ignorant of the technological limitations, quirks and caveats associated with broadband access platforms. For example, it is extraordinarily easy to overestimate the provision of broadband services using published information from the National Broadband Map (http://www.broadbandmap.gov/) if one does not understand how these technologies actually function at the local level or how these data are reported. From the distance constraints associated with xDSL to the wireless coverage issues associated with cellular technologies, technological constraints matter.

We hope this chapter provides analysts with a gentle introduction to broadband access technologies and their associated infrastructure systems – along with a starting point from which to consider how the nature of broadband infrastructure may impact large-scale inquiries concerning provision, access, equity, policy and economic development within a region. These facets are strongly co-dependent and deserve a more thorough integration in future work.

Notes

1 There are also Tier 3 and 4 providers in the U.S., but they do not maintain direct access to the global backbone and classifying them as WANs can be inaccurate. All network services at this level are purchased through Tier 1 or 2 carriers and resold. Windstream and LightPath are good examples of Tier 3 carriers.

2 Integra also serves Seattle, WA, Las Vegas, NV, Salt Lake City, UT and a variety of other major cities in the Western United States.

3 Repeaters are devices placed on long runs of coaxial cable to help retransmit signals and extend their geographic range.

4 For those familiar with basic networking architecture, cable systems correspond to a classic bus topology in neighborhoods, but adhere more strongly to a star topology at the fiber node level and a ring topology within the fiber loops of the HFC system.

5 Channel bonding via DOCSIS 3.0 allows cable operators to bond multiple radio frequency channels together between the headend and end-user premises. This can

dramatically increase both upstream and downstream transmission performance, with bandwidth reaching 120 Mbps and 160 Mbps, respectively (Gorshe et al., 2014).

6 ATM is a mature technology that provides functionality that is similar to both circuit switching (similar to POTS) and packet switching.

7 For example, the JDS Uniphase TB4-DIS-QUAD-S-T-berd 4000 Quad Otdr Basic Package, W/sc Conn Light Source retails for $10,805.95 (USD). Smaller handheld devices retail anywhere from $2,000 to $6,000 USD (see http://tinyurl.com/od9t5oe).

8 Handoffs refer to the process by which cellular phone conversations are passed from one cell to another (Newton, 2002).

9 SIMs are integrated circuits that store the international mobile subscriber identity and its key to authenticate subscribers for network use (Newton, 2002).

10 Mobile data plans can be extremely expensive. For example, in May 2014, the $75 USD single line plan from Verizon provided for 2 GB of data, unlimited voice and text, but charged $15 USD for every 1 GB overage. A 50 GB package (Verizon's largest) costs $375 USD per month.

References

Abe, G. (2000). *Residential broadband.* Cisco Press.

Abichar, Z., Peng, Y., & Chang, J. M. (2006). WiMAX: The emergence of wireless broadband. *IT Professional*, 8(4), 44–48.

Agashe, P., Rezaiifar, R., & Bender, P. (2004). cdma2000® high rate broadcast packet data air interface design. IEEE Communications Magazine, 42(2), 83–89.

Agrawal, D., & Zeng, Q.A. (2015). *Introduction to wireless and mobile systems.* Cengage Learning.

Alcaraz, C., Fernandez, G., & Carvajal, F. (2012). Security aspects of SCADA and DCS environments. In J. Lopez, R. Setola, & S.D. Wolthusen (Eds.), *Critical infrastructure protection* (pp. 120-149). Berlin, Heidelberg: Springer.

Bloomberg. (2001). *Rats! The cable is down again.* Retrieved from http://www.businessweek.com/stories/2001-07-11/rats-the-cable-is-down-again

Bonomo, J. (2012). *Bringing fiber-optic connectivity to lower Manhattan post-hurricane Sandy.* Verizon. Retrieved from http://newscenter.verizon.com/residential/news-articles/2012/12-lower-manhattan-nyc-sandy-restoration/

Brown, M. (2011). *Super Bowls and Royal Weddings – Betting on a sure thing.* Cisco. Retrieved from http://tinyurl.com/oogrvm6

Cartwright, J. (2000). *Introduction to load coils and bridge taps.* Retrieved from URL: http://www.info-ed.com/support/connection_support.htm

Cassidy, J. (2014). We need real competition, not a cable-Internet monopoly. *The New Yorker.* Retrieved from URL: http://tinyurl.com/o346s36

Courtney, M. (2013). Whatever happened to broadband over power line? *Engineering and Technology* Magazine. Retrieved from http://eandt.theiet.org/magazine/2013/10/broadband-over-power-line.cfm

Dahlman, E., Parkvall, S., & Skold, J. (2013). *4G: LTE/LTE-advanced for mobile broadband.* Academic Press.

Federal Communications Commission [FCC]. (2013). *Measuring broadband America.* Retrieved from http://www.fcc.gov/measuring-broadband-america/2013/February

Fischer, C.S. (1992). *America calling: A social history of the telephone to 1940.* University of California Press.

Gabel, D., & Kwan, F. (2001). Accessibility of broadband telecommunication services by various segments of the American population. In B.M. Compaine & S. Greenstein (Eds), *Communications policy in transition: The Internet and beyond* (pp. 295–320). Cambridge: MIT Press.

Goleniewski, L. (2002). *Telecommunications essentials: The complete global source for communications fundamentals, data networking and the Internet, and next-generation networks.* Addison-Wesley Professional.

Gorshe, S., Raghavan, A., Starr, T., & Galli, S. (2014). *Broadband access: Wireline and wireless – alternatives for Internet services.* New York: Wiley.

Graziano, D. (2013). AT&T LTE network tops speed test, but Verizon still has the best LTE coverage. BGR. Retrieved from http://bgr.com/2013/03/11/4g-network-speeds-368339/

Green Jr, P.E. (2002). Paving the last mile with glass. *IEEE Spectrum, 39*(12), 13–14.

Grubesic, T.H. (2008). Spatial data constraints: Implications for measuring broadband. *Telecommunications Policy, 32*(7), 490–502.

Grubesic, T.H. (2012a). The US national broadband map: Data limitations and implications. *Telecommunications Policy, 36*(2), 113–126.

Grubesic, T.H. (2012b). The wireless abyss: deconstructing the US National Broadband Map. *Government Information Quarterly, 29*(4), 532–542.

Grubesic, T.H. (2015). The broadband provision tensor. *Growth and Change.* DOI: 10. 1111/grow.12083

Grubesic, T.H., & Horner, M.W. (2006). Deconstructing the divide: extending broadband xDSL services to the periphery. *Environment and Planning B: Planning and Design, 33*(5), 685.

Grubesic, T.H., & Murray, A.T. (2002). Constructing the divide: Spatial disparities in broadband access. *Papers in Regional Science, 81*(2), 197–221.

Grubesic, T.H., Matisziw, T.C., & Murray, A. (2010). Market coverage and service quality in digital subscriber lines infrastructure planning. *International Regional Science Review, 34*(3), 368–390.

Grubesic, T.H., Matisziw, T.C., & Ripley, D.A. (2011). Approximating the geographical characteristics of Internet activity. *Journal of Urban Technology, 18*(1), 51–71.

Higginbotham, S. (2010). *Mobile milestone: Data surpasses voice traffic.* Gigaom. Retrieved from http://gigaom.com/2010/03/24/mobile-milestone-data-surpasses-voice-traffic/

Hogg, D.C., & Chu, T.S. (1975). The role of rain in satellite communications. *Proceedings of the IEEE, 63*(9), 1308–1331.

Horrigan, J.B., Stolp, C., & Wilson, R.H. (2006). Broadband utilization in space: Effects of population and economic structure. *The Information Society, 22*(5), 341–354.

IHS Technology. (2013). *Broadband Internet penetration deepens in US; cable is king.* Retrieved from https://technology.ihs.com/468148/

Integra. (2014). *Integra network map.* Retrieved from http://www.integratelecom.com/pages/network-map.aspx

JD Power. (2013). *Performance and reliability problems decline in both residential TV and Internet services; quality and connection speeds continue to improve.* Retrieved from http://tinyurl.com/oz6869f

Kazovsky, L.G., Cheng, N., Shaw, W-T, Gutierrez, D, & Wong, S-W. (2011). *Broadband optical access networks.* New York: Wiley.

Lee, G., & Murray, A.T. (2010). Maximal covering with network survivability requirements in wireless mesh networks. *Computers, Environment and Urban Systems, 34*(1), 49–57.

Mack, E.A., & Grubesic, T.H. (2009). Forecasting broadband provision. *Information Economics and Policy, 21*(4), 297–311.

Malecki, E.J., & Wei, H. (2009). A wired world: the evolving geography of submarine cables and the shift to Asia. *Annals of the Association of American Geographers, 99*(2), 360–382.

Mattern, F., & Floerkemeier, C. (2010). From the Internet of computers to the Internet of things. In K. Sachs, I. Petrov, & P. Guerrero (Eds.), *From active data management to event-based systems and more* (pp. 242–259). Berlin, Heidelberg: Springer.

Moyer, M. (2009). The everything TV. *Scientific American, 301*(5), 74–79.

Murray, A.T., & Grubesic, T.H. (2013). Fortifying large scale, geospatial networks: Implications for supervisory control. In *Crisis Management: Concepts, Methodologies, Tools, and Applications* (pp. 224–246) (3rd ed.). Hershey: Information Science Reference.

Nafea, H.B., Zaki, F.W., & Moustafa, H.E. (2013). Performance and capacity evaluation for mobile WiMAX IEEE 802.16 m Standard. *Nature, 1*(1), 12–19.

Newton, H. (2002). *Newton's telecom dictionary.* New York: CMP Books.

Oden, M., & Strover, S. (2002). *Links to the future: The role of information and telecommunications technology in Appalachian economic development.* Appalachian Regional Commission. Retrieved from http://eric.ed.gov/?id=ED467710

O'Kelly, M.E., & Grubesic, T.H. (2002). Backbone topology, access, and the commercial Internet, 1997–2000. *Environment and Planning B, 29*(4), 533–552.

Organisation for Economic Co-operation and Development [OECD]. (2013). *OECD broadband portal.* Retrieved from http://www.oecd.org/sti/broadband/oecdbroadband-portal.htm

PeeringDB. (2014). *Peering database.* Retrieved from http://tinyurl.com/p83qohr.

Prieger, J.E., & Hu, W.M. (2008). The broadband digital divide and the nexus of race, competition, and quality. *Information Economics and Policy*, 20(2), 150–167.

Schilke, O., & Wirtz, B.W. (2012). Consumer acceptance of service bundles: An empirical investigation in the context of broadband triple play. *Information & Management, 49*(2), 81–88.

Shillington, L., & Tong, D. (2011). Maximizing wireless mesh network coverage. *International Regional Science Review, 34*(4), 419–437.

Tauri Group. (2013). *State of the satellite industry report, June 2013.* Retrieved from http://www.sia.org/wp-content/uploads/2013/06/2013_SSIR_Final.pdf

Ulm, J., & Weeks, B. (2007, June). Next play evolution: beyond triple play & quad play. In *IEEE International Symposium on Consumer Electronics, 2007.* ISCE 2007 (pp. 1–6). IEEE.

Vaughan-Nichols, S.J. (2004). Achieving wireless broadband with WiMax. *Computer, 37*(6), 10–13.

ViaSat. (2014). *High speed satellite Internet from space.* Retrieved from http://www.exede.com/

4   Broadband data

Broadband data in the United States has a relatively short, but sordid history. Their roots can be traced back to the Federal Communications Commission (FCC) and the Form 477 program. Established in 2000, the goal of Form 477 was to provide the FCC with a suite of uniform and reliable data concerning broadband services, local telephone competition and mobile telephony (FCC, 2013). Today, broadband data is available in many flavors, including that from the FCC Form 477, the National Broadband Map (NBM; NTIA, 2014a), state agencies (IN.gov, 2014), cities (NYC, 2014) and private corporations (Ookla, 2014; Verizon, 2014).

Given the wide range of data available, a continuing challenge for analysts, policymakers and consumers of broadband is making sense of what the various data represent. In particular, there are serious implications concerning the ways in which broadband data are collected, aggregated, summarized and visualized for empirical analysis and policy development. The purpose of this chapter is three-fold. First, it provides readers with a candid overview of the strengths, weaknesses and uncertainties associated with a variety of currently available broadband data, including that from the NBM. This is important because the mythology associated with the NBM is one of comprehensiveness and completeness (Grubesic, 2012a). Second, to add depth and clarity, this chapter also provides an overview of the challenges associated with using legacy FCC Form 477 data, from 1999 to 2010. This is important because the Form 477 filings were the only data source available on broadband deployment for over a decade and remain important when conducting longitudinal or time-series analyses. Finally, because broadband is inherently spatial, this chapter provides readers with a primer on the geography of broadband and the various administrative units used to collect, aggregate and synthesize broadband data. Not only is this fundamental to understanding how services are provisioned, it is also critical to better appreciate how elements of uncertainty creep into empirical analysis and telecommunications policy development.

FCC Form 477

December 1999 – December 2004

Sometimes it is easier to start at the beginning. Although the NBM and its associated data have certainly generated much excitement (and use) since its introduction in 2011, FCC Form 477 data predate the NBM by more than a decade. As noted earlier in this chapter, Form 477 data provide one of the few sources of long-range, time-series data on broadband deployment for the United States. Established in 2000, the goal of Form 477 was to provide the FCC with a suite of uniform and reliable data concerning broadband services, local telephone competition and mobile telephony services (FCC, 2013).[1] Fifteen years later (2015), the FCC still collects these data, requiring all facilities-based providers in the U.S. to file a Form 477 twice a year.[2] Facilities-based providers typically include incumbent local exchange carriers (ILECs) and competitive local exchange carriers (CLECs), cable operators, terrestrial fixed wireless providers, electric utilities, municipalities that own/operate/provide broadband to subscribers within a region (FCC, 2014).

One of the interesting quirks with early tabulations of the Form 477 data (December 1999 – December 2004) is that only facilities-based providers with at least 250 or more broadband lines in a state were required to submit information about their services. Data were collected at the ZIP code level and published on the FCC's Wireline Competition Bureau website. The final product was a total count of broadband providers within each ZIP code for the entire U.S.[3]

The early Form 477 data collection strategy and subsequent structure had six major impacts on empirical efforts and telecommunications policy development, all of which introduced significant elements of uncertainty into the broadband data. First, it systematically discounted and/or ignored small, competitive providers (< 250 lines) that served a highly localized customer base. These providers were primarily located in rural and remote areas, so the resulting tabulations of the Form 477 data displayed a modest urban bias. Second, for ZIP code areas with three or fewer providers, information about the exact number of broadband suppliers was suppressed. At the time, providers were concerned that the disclosure of such information would place them at a competitive disadvantage (Grubesic & Murray, 2004). As a result, it was impossible for analysts to evaluate the presence of monopolies or duopolies in early broadband markets. Third, no information regarding type of platforms deployed, price, speed or customer counts for a ZIP code area was made available. As a result, evaluating quality of service (QoS) and platform-based competition was not possible. Fourth, the selection of ZIP codes for reporting purposes, while convenient, is highly problematic. ZIP codes do not actually correspond to geographic areas (Grubesic, 2008). ZIP codes are developed by the United States Postal Service (USPS) to optimize the delivery of mail. Thus, they are structured as address ranges along street networks. ZIP code areas are inexact interpolations of these address ranges (Grubesic & Matisziw, 2006). ZIP code areas are *highly* variable in size and dynamic over time, changing structure at the whim of the USPS (Grubesic, 2008). For example, there are several ZIP code areas located in the Intermountain West that are thousands of square miles in size, while others in large metropolitan areas like Philadelphia are less than one square mile in size, or smaller. These size differentials make comparisons difficult. A fifth issue associated with the early Form 477 data was the embedded assumption that if a ZIP code area reported the presence of at least one broadband provider, it was ubiquitously available throughout the ZIP code area. This was (and remains) a poor assumption for any area, given that spatial heterogeneity is a common feature in the distribution of many services, including broadband (Strover, 2001; Grubesic, 2003; Grubesic & Murray, 2004; Mack & Grubesic, 2009). Further, as detailed in Chapter 3, many broadband platforms are geographically limited. This includes xDSL. As a result, for many ZIP code areas, although xDSL service is provisioned, its actual availability is limited to households or businesses within the wire center service boundary associated with a central office (Grubesic & Murray, 2005; Grubesic, 2008).[4]

Consider, for example, Figure 4.1, which illustrates Nye County, Nevada, and ZIP code 89049 (Tonopah, NV). Nye County is the third largest county in the contiguous U.S., and spans 18,159 square miles (47,030 km²), roughly equivalent to the combined size of New Jersey and New Hampshire. Within Nye County is the Tonopah ZIP code area (89049), which had a population of 3,249 in 2010. As detailed by Grubesic (2008), this region was not an extremely large broadband market in 2004, but it was a geographically expansive one. According to the 2004 FCC Form 477 database, there were between one and three broadband providers in this ZIP code.[5] However, the FCC did not disclose the location of these providers. If one assumes that xDSL was one of the available technologies in this ZIP code area during 2004, understanding where a provider offered service would be an important consideration. The potential location for xDSL service provision can be narrowed to the region proximal to the Tonopah central office (CO), which is located on the western border of 89049. As detailed in Chapter 3, one could also assume that xDSL availability would extend about 18,000 ft. from the CO location in Tonopah, but no further. In short, terrestrial broadband provision was extremely limited in 89049.

This example clearly captures the uncertainty embedded within FCC Form 477 data. To summarize:

1   These data only capture information from providers with 250 or more lines provisioned.
2   The actual number of providers is unknown (somewhere between 1 and 3).
3   There is no information on pricing, platforms, speed or customer counts.
4   Data is summarized within a ZIP code area, which in this case, is extremely large.
5   Form 477 summary statistics at the ZIP code level are seriously misleading because of the known effects of spatial heterogeneity.
6   xDSL service, if available in Tonopah, is not available throughout the 89049 ZIP code area due to the geographic constraints of xDSL technology.

June 2005 – June 2008

In time, the collection procedures for the FCC Form 477 data were amended. Starting in 2005, providers that operated fewer than 250 lines within a state were also included in the Form 477 data. That said, the remaining data collection

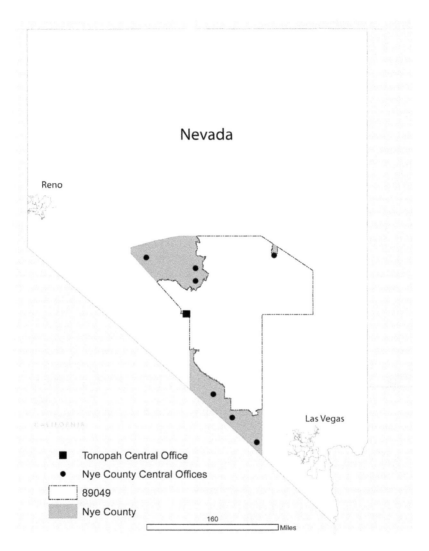

Figure 4.1 xDSL coverage asymmetries, Nye County, Nevada.

procedures and reporting requirements were identical to 1999–2004. This small change, however, had major impacts on the broadband landscape. First, dropping the 250-high speed line reporting threshold made a more complete picture of broadband provision possible for the U.S. Again, for the first time, small providers, many of which were located in the rural and remote areas, were included in the tabulations. To some degree, this helped mediate the underlying urban bias in the 1999–2004 data. This change also meant that comparisons

between the first iteration of Form 477 data (1999–2004) and the second iteration (2005–2008) were impossible. Although a time series does exist, the data
are asymmetric.

2009 – present

The last major change to the FCC Form 477 database occurred in December 2008.
Most notably, ZIP codes were dropped as the tabulation units in favor of Census
tracts. This is a major improvement because Census-based area units are nested
spatial units that are standardized (at least where population is concerned) by the
Census Bureau. So, even though there is some variation in unit size, it is less
frequent and of less magnitude for tracts when compared to ZIP codes. Tracts
are also more consistent over time. ZIP codes are updated monthly (Grubesic,
2008), but Census tracts are restructured every ten years for the decennial Census.
This added level of spatiotemporal stability helps minimize uncertainty, even if
the caveats associated with spatial heterogeneity of broadband provision remain.
A second major change to the Form 477 data during this period was the inclusion
of speed-tier information.[6] In theory, this is a major advancement because it provides a snapshot of how fast delivered services might be for a region. This is the
first QoS metric to ever be included in the Form 477 data. As detailed by Mack
and Grubesic (2014), despite these changes, two persistent issues remain. There is
still no information about the types of broadband platforms available at the tract
level. In other words, the data establish the presence of broadband, but no information is available that details if services are delivered by cable, xDSL, fiber, satellite or something else. Further, the speed tiers are not particularly informative. It
is important to remember that these data represent "advertised speeds" as reported
by the providers. There are no actual metrics associated with "realized speeds" as
reported by the consumer (Grubesic, 2012a, 2015). This issue, coupled with the
low speed threshold of 200 kbps for broadband set by the FCC, drastically limits
the ability for analysts to determine QoS, especially where speeds/bandwidth are
concerned. Finally, it is important to note that the tract information reported by
the FCC were initially based on year 2000 Census tracts, which makes integrating
2010 Census data incredibly difficult.

Proceed with caution

Problems abound with the FCC Form 477 data. Although they remain the one and
only viable link to recent broadband provision history, users must proceed with
caution. Of the three generations of data, the 2009 to present version provides the
highest spatial resolution, includes the most carriers, the underlying geographic
tabulation unit is the most consistent, but there is an extremely unreliable metric associated with speed reported (advertised). Unfortunately, these data cannot
be combined with previous iterations of the Form 477 data because ZIP codes
and census tracts are largely incompatible. Worse, even the earliest versions of
the data (2000–2004 and 2005–2008) are incompatible because of the changes in

reporting rules. In sum, these details are not meant to "scare" readers away from the data or their use, quite the opposite is true. Much can be learned from these data, even with their known imperfections. Users must simply be aware of the caveats associated with their use so that common mistakes can be avoided and meaningful empirical results can be generated.

The National Broadband Map

The NBM was released in February 2011. It was a joint effort between the FCC and the National Telecommunications and Information Administration (NTIA), which is an agency within the U.S. Department of Commerce. Unlike previous iterations of broadband data, including the FCC Form 477 compilations, the NBM provided the first searchable map of broadband provision, for the entire U.S.

The roots of the NBM are tied to the Broadband Data Improvement Act of 2008 (BDIA, 2008). The BDIA established a number of initiatives to improve the character of state and federal broadband data. Better metrics concerning availability, location and QoS were the primary targets. In July 2009, the NTIA responded to the BDIA's call for better data and announced a formal mechanism to fund the State Broadband Data and Development Program (SBDDP). Specifically, a pool of funds was made available to the NTIA to start the broadband mapping process. These funds were drawn from the Broadband Technology Opportunities Program (BTOP) which was funded as part of the American Recovery and Reinvestment Act of 2009 (Kruger, 2009). Approximately $350 million were allocated to the NTIA for distribution to the SBDDP. To maintain consistency and regularity in data collection efforts, a single entity in each state was selected and charged with acquiring, processing and reporting broadband data to the NTIA (ibid.). For example, in the State of Ohio, the Ohio Office of Information Technology was initially awarded $500,000 for planning the data collection efforts and $1.3 million for collecting the data. In Oregon, the Oregon Public Utilities Commission was awarded $500,000 for planning and $1.6 million for data collection. Both states subsequently hired subcontractors for this effort.

The problem with the SBDDP, and ultimately the NBM, is that data collection proved to be more difficult than anticipated. Although the NTIA provided a template of what data needed to be collected and how it should be processed prior to its conveyance to the NTIA for mapping, state agencies found that building relationships and interfacing with broadband providers was somewhat difficult. For example, the first iteration of the NBM released in 2011 had extremely poor provider participation. IDInsight (2011) tracked participation rates for broadband providers in each state and the results were not good. For instance, 14 states had provider participation rates under 60%.[7] If that was not alarming enough, only 27% of the providers in the state of Virginia participated. Unfortunately, this did not stop the NTIA from uploading and reporting the incomplete results on the NBM. Worse, the details concerning provider participation were buried on the NBM site and most users of these data were largely unaware of the errors embedded in the NBM until these uncertainties were highlighted by third-party efforts (IDInsight, 2011; Lennett & Meinrath, 2011; Meinrath, 2011; Grubesic, 2012a, 2012b).

At the time this chapter was written, the most current iteration of the NBM data was December 2013, and these data were live on the NBM site (http://broad bandmap.gov/). Provider reporting problems remained. For example, Connected Nation (2014), a subcontractor and the designated broadband data collection entity for the state of Texas reported the following:

1   The December 2013 data update submission included datasets for 93.95% of the Texas provider community, or 202 of 215 providers.
2   Of the 202 providers for which data were submitted, 184 actively participated in the update process, but 18 additional non-participating providers had their coverage areas estimated by Connected Nation.
3   Of the 184 providers that did actively participate, 108 supplied an update and 66 reported no change.
4   10 additional providers that previously supplied data were non-responsive, so Connected Nation used previously submitted data.

Again, although it is not our intention to alarm readers, a careful inspection of these disclosures from Connected Nation (a.k.a., Connected Texas) are cause for concern. First, the data are incomplete because not all providers participated. Second, portions of the data reported to NTIA are estimates of provider coverage derived by Connected Texas. There are no details on how these estimates were generated. Third, some of the data reported and labeled as December 2013 are actually legacy data from previous collection efforts (perhaps June 2013 or earlier), because providers were non-responsive.

For readers, the takeaway from this brief tour of NBM data for Texas is a simple one. Texas is not alone. All 50 states, Puerto Rico and Washington, DC suffer from many of the same data uncertainty issues. The grantee methodologies (NTIA, 2013a) are a treasure trove of information concerning errors, estimates and related problems with the NBM data. Moreover, although the broadband data collection entities try to follow the data model and reporting template required by NTIA, there is significant variation between states. A careful inspection of these methodologies over the various iterations of the NBM reveals that some states have changed their approach between collection periods – generating internal inconsistencies over time. Of course, all of these uncertainties are eventually absorbed and reflected by the NBM.

In the following subsections, we provide readers with an overview of the strengths and weaknesses associated with the NBM. This is important for several reasons. First, the NBM currently functions as the de facto source for information on broadband provision for the United States. Thus, developing a deeper understanding of what the data are and how they can be used is critical for conducting empirical analysis, especially if the outcomes of these analyses are tied to policy development. Second, spatial data are special (Anselin, 1989). Although the NBM gives the illusion of accuracy, completeness and data portability, the collected broadband data have a strong spatial component and are intimately tied to the local geometries of administrative units.[8] As a result, a misunderstanding of these reporting procedures, especially where data aggregation is concerned, can lead to significant systemic biases in representation and interpretation. Finally, care is taken not to repeat the level of diagnostic detail provided in Grubesic (2012a), which discusses the nuances of wireline NBM data and Grubesic (2012b), which does the same for wireless NBM data. Readers are encouraged to consult these references for all the details associated with data uncertainty in the NBM and methods to mitigate it.

It is also important to mention the many facets of the NBM that we do not cover in the following sections. As detailed previously, because we are focusing on terrestrial broadband provision, wireless coverage is not explored in this chapter. Likewise, issues associated with community anchor institutions (CAIs) and middle mile infrastructure points are also left for future work. Each of these databases has their own problems and space constraints limit us from addressing them in a meaningful way. Further, we believe that the bulk of NBM consumers, analysts and policymakers are primarily interested in terrestrial broadband provision at the block level, so this is where we focus our efforts. In short, the following subsections are an attempt to identify "weeds" in the broadband data ecosystem and to provide consumers of these data some additional perspective on why circumspection is required for the use of NBM data.

A broad brush

At the time of its release, Steven Rosenberg, the chief data officer with the FCC's Wireline Competition Bureau, touted the NBM as the "largest and most detailed map of broadband ever created" (Meinrath, 2011). He was correct. However, what Rosenberg failed to mention is that bigger does not necessarily mean better. As detailed earlier, provider participation rates for the first iteration of the NBM were low (IDInsight, 2011), thereby creating an incomplete snapshot of broadband provision. In sum, approximately 74% of all providers participated nationally (3,400 of 4,600; NTIA, 2011a).[9]

The NBM has several additional problems worth noting. First, the NBM relies on self-reported provider data. To be clear, the entity charged with collecting broadband data in a state (e.g., Connected Nation) is required to survey broadband providers to ascertain where they are offering broadband services. Participating providers transmit this information back to the data collection entity, where it is injected into the NTIA (2011b) data model (locally) and prepared to upload to the NTIA for secondary processing and connection to the NBM. The problem with this type of reporting process is that there was very little oversight on the providers and the validation procedures that were used can be weak and/or poorly documented. In Texas, for example, reported wireline provision data were validated using third-party data sources, but no explicit information on how this information was used is provided in the technical notes.[10] In other states, including Oregon, confidence values were assigned to the reported broadband data.

These values ranged from 0 (coverage area has not been reviewed) and 10 (extremely low, single source quality control) to 100 (perfect, multiple validation\ verification sources, with complete alignment with sources and ground truth verification activities). This information was not provided with the NBM, in any way.

A second problem with the reported data is that providers often "paint their coverage areas with a broad brush" (Lennett & Meinrath, 2011, 2). Ironically, it is the NTIA that allows providers to do this. As detailed by Grubesic (2012a), there were many instances in the first generation NBM map where providers reported that broadband was available, but these locations, in fact, did not have broadband service. The reason for this discrepancy is that providers were asked to identify locations where broadband services were already active, *and* locations that they could service within 7–10 days. There is a huge difference between reported locations that have broadband and those that *could* have broadband within 7 to 10 days. This certainly facilitates painting with a broad brush and more technical details on why this is a problem are discussed later in this chapter.

Third, the NBM merges the reporting of business and residential services. Thus, although some areas appear to have a robust assortment of providers and high levels of competition, the majority of providers within the area may be targeting businesses, not residential consumers. Again, this is misleading and the data should be characterized accordingly. More details on how such reporting quirks manifest locally are covered in Chapter 6 of this book.

The nuts and bolts of wireline NBM data

As highlighted previously, the core data elements of the NBM are associated with broadband provision. In particular, the data are focused on two central variables: 1) the number of providers in an area and 2) the advertised download/upload speeds of the providers. The omission of a pricing variable is a clear and inexcusable weakness of the NBM, but for now, provider count and bandwidth will suffice for our discussion. Both metrics are tied to Census-defined tabulation units known as blocks. Census blocks are the smallest administrative unit for which the U.S. Census Bureau collects and reports decennial survey information. In 2000, there were 8,262,363 unique Census blocks for the United States and Puerto Rico. By 2010, this number had grown to 11,155,486 (Census, 2000a, 2010). Blocks are bounded by streets, streams, highways and railways and can be impacted by larger cultural features such as secondary schools, prisons and college campuses. Figure 4.2 displays a fairly typical tessellation of block features in a large urban area. In this case, the blocks are located in Philadelphia, PA. It is important to note that block geometries are somewhat regular within residential areas (e.g., rectangular), but highly irregular near industrial areas (e.g., Delaware River waterfront) and in areas proximal to major transportation infrastructure (e.g., Interstate 95).

There are many advantages to using Census administrative entities for tabulating broadband provision statistics. First, the Census uses hierarchically nested tabulation units. Ultimately, this makes the process of data aggregation easier.

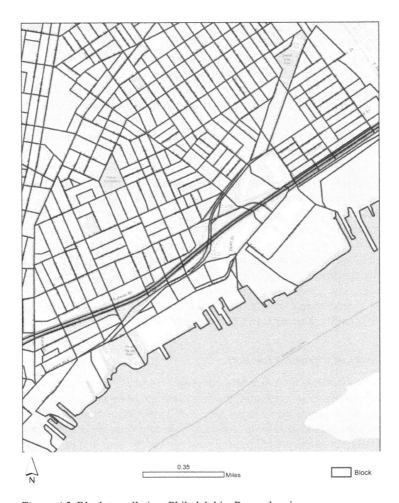

Figure 4.2 Block tessellation, Philadelphia, Pennsylvania.

For example, a Census block is always unique to, and can never cross the boundaries of Census tracts or block number areas (Census, 2000b). Thus, when aggregating data, one can be reasonably confident that there are no overlapping boundary issues that may lead to double-counting of features. Second, it is important to reiterate that although major urban areas in the U.S. maintain a high-density grid of relatively small and compact blocks, more recent suburban and exurban developments display curvilinear street patterns, cul-de-sacs and larger expanses of parks, greenways or other spaces. In turn, the Census blocks in these areas reflect a similar structure. Consider the blocks highlighted in Figure 4.3, located in/around Bloomington, Indiana. The downtown core of Bloomington displays

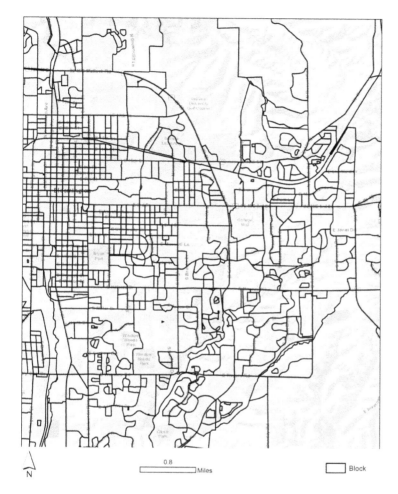

Figure 4.3 Block tessellation, Bloomington, Indiana.

highly regularized blocks, with a size and shape similar to the blocks highlighted for Philadelphia. However, the residential developments found in the southeastern portions of Bloomington are more irregular and highly curvilinear. This pattern is repeated throughout virtually every metropolitan area in the United States as one moves away from urban cores to suburban and exurban locations. Moreover, as one moves further from the residential developments on the edge of the city, it is important to notice that the blocks become much larger and correspond to agricultural areas or undeveloped, forested tracts of land. Again, this pattern is repeated throughout the U.S.

Figure 4.4 Broadband geographies, Travis County, Texas.

Block geographies are important for discussing the NBM because the NTIA collects wireline data in two different ways. For blocks that are *less than* two square miles in size, broadband providers are required to report block ID numbers where broadband service is available. For blocks *larger than* two square miles in size, providers are required to collect and submit either address data or road segment data where broadband is available. Consider Figure 4.4a, which highlights the native Census tessellation of both large and small blocks for Travis County, Texas in 2010. As one would expect, the blocks in more rural areas, located outside of the Austin city limits are larger in spatial extent. Figure 4.4b displays all of the small blocks in Travis County where broadband is available and all of the road segments associated with large blocks where broadband is available for 2013

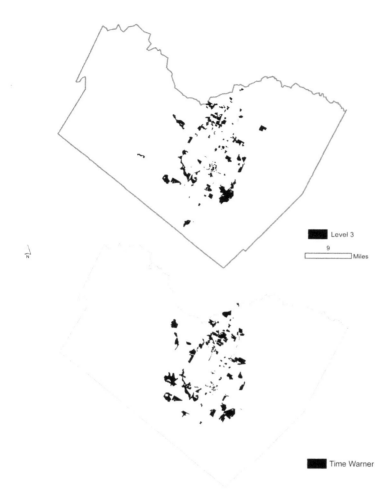

Figure 4.5 Optical fiber providers, Travis County, Texas.

(NTIA, 2013b). One important nuance to mention is the combination of regularly and irregularly shaped areas depicted in the color white for Figure 4.4b. These correspond to areas where broadband is not available. The irregular areas primarily correspond to water bodies, such as Lake Travis (northwest) and Town Lake (i.e., Colorado River), which runs through the heart of Austin.

NBM post-processing

As mentioned earlier, Connected Texas is charged with surveying, tabulating and reporting the broadband availability data from providers throughout the state and conveying these data to NTIA for processing and injection to the NBM.[11] In its simplest form, Figure 4.5 highlights the results of post-processing for two broadband providers. Specifically, for blocks less than 2 square miles, Figure 4.5a highlights the spatial distribution of optical fiber supplied by Level 3 Communications for Travis County. Figure 4.5b highlights optical fiber provision by Time Warner for Travis County. Although portions of these service areas and their associated blocks overlap, they are not identical. In a nutshell, this is what the NBM is. A large tabulation of provider-specific coverages.

For Travis County, there are four additional providers that offer a variety of different broadband platforms for the area, including Verizon, Time Warner (cable), MegaPath and Grande Communications. Unfortunately, rather than simply providing a single coverage of blocks that reports broadband provision and a unique count of providers, the NBM assigns a coverage to each provider, but these coverages must be extracted individually from the shapefile provided by the NTIA. For those not intimately familiar with geographic information systems, perhaps the simplest way to communicate this jumble of data is as follows. In 2010, Travis County had a total of 15,922 Census blocks. Ideally, each one of these blocks should have a count of unique providers, their associated platforms (e.g., cable, xDSL) and information on broadband speed(s). Instead, for Travis County, the NBM provides 53,052 blocks with this information. In other words, each block is reported multiple times if it is served by a provider. Not only does this drastically inflate the database, it makes processing these data more time consuming.[12]

In addition to reporting the basic spatial footprint of broadband provision, as detailed above, a second major task undertaken by the NTIA is to calculate broadband availability statistics. The NTIA (2014b, 1) identifies these post-processing tasks as important, because:

> This analysis transforms address and road segment data into consistent counterpart data of blocks in order to calculate broadband availability statistics. This allows us to develop population and housing statistics across the entire dataset.

The NTIA (2014b, 1) notes that there are several challenges with this process:

> For example, given just individual road segment data, it is not possible for us to determine the number of households on that particular road segment. As a result, it is impossible to find the number of households with broadband availability.

This, of course, is where additional elements of uncertainty begin to creep into the NBM data. In an effort to estimate the number of households that *might* be covered by reported broadband service within a block, the following logic (and associated geoprocessing) is used (NTIA, 2014b, 1):

The calculation first estimates the point locations of households in census blocks greater than two square miles given the number of these households and the distribution of roads within that block. Our model follows the basic premise that the distribution of households exactly follows the distribution of roads. We therefore grow household along the distribution of roads.

This is a poor assumption for several reasons. First, there are no details provided by the NTIA concerning how the households "grow" along the distribution of roads. Is this some type of random assignment? Does it adhere to a statistical distribution that corresponds to settlement density within the block? Second, although settlement patterns do often correspond to road networks, there is nothing "exact" about it (Xie, 1995; Mrozinski & Cromley, 1999; Reibel & Bufalino, 2005; Matisziw, Grubesic, & Wei, 2008). This is especially true in rural areas, where there are no guarantees that households or businesses are tightly co-located with the local street network. In many instances, there are large offsets between road segments and associated premises.

Next, in an effort to fairly tabulate the distribution of households and generate unbiased wireline statistics, the following rules are applied (NTIA, 2014b, 1):

Once we have this distribution, using geospatial operations, we find the number of modeled household points that are within 500 feet of every broadband availability location (road segment and address point) and divide by the total number of households within that block. This gives us a distribution of households and population within that block. For the purposes of this calculation, *when a provider offers service within a block that is less than two square miles, we assume that the provider is offering service throughout the block.* Once complete, we then combine the random point results and the blocks less than two square miles results into one table for consistent wireline statistics. [emphasis added]

Once again, this logic is faulty and allows for significant spatial data uncertainty to enter the statistical calculations. For example, consider the 500-foot buffer applied to each road segment that is purported to have broadband availability in the reported NBM data. Figure 4.6 highlights exactly how erroneous that assumption can be. Highlighted is a house that is approximately 750 ft. (~230 m) from the buffered road segment. Given the somewhat ambiguous processing rules applied by the NTIA, it is unclear if this household would be considered "served" or "unserved" in the NBM. This problem is repeated thousands of times in rural areas throughout Texas, and likely many more times throughout the rural portions of the U.S. Ironically, the notion of proximity to roadways and household offsets have been addressed in the geocoding literature (Goldberg, Wilson, & Knoblock, 2007; Zandbergen, 2009; Murray, Grubesic, Wei, & Mack, 2011) and there is a fairly large literature discussing strategies to mitigate such error. Offset distances are highly variable and are largely contingent upon the local spatial structure of

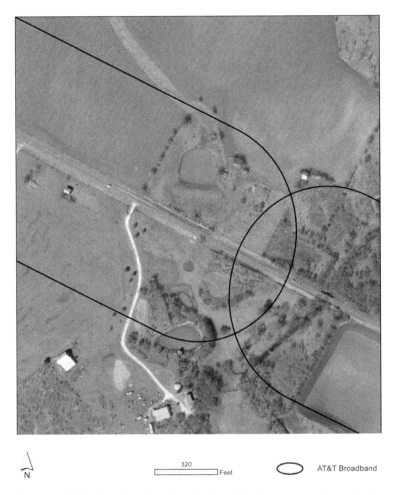

Figure 4.6 500-ft buffers for estimating broadband coverage in large blocks.

a region. In major cities, offset distances are small. In suburban areas, they are somewhat larger. In rural areas, as illustrated by Figure 4.6, they are large. The use of a single distance criterion (500 ft. [~153 m]) for the entire United States for calculating broadband coverage is a poor choice. It should be driven by local spatial structure, not convenience.

Uncertainty in broadband coverage

The flaws in logic and associated reporting conventions detailed above, manifest concretely in the NBM, as do divergent broadband service calculations between states. Although space constraints limit us from providing a close examination of each problem, we highlight several of the most salient issues and provide examples in this subsection.

One of the most significant problems with error and uncertainty in the NBM are the divergent broadband service calculations that are prevalent between states. As detailed in Chapter 3, xDSL coverage is distance constrained. Service is generally available to a household or business if it is located within 18,000 ft. (5.5 km) of a CO. The problem with the NBM is that state methodologies for calculating xDSL coverage varies widely. Drawing from the June 2011 grantee methodology statements, Table 4.1 illustrates a small sample of this methodological variation

Table 4.1 Sampling of state methodologies for calculating xDSL coverage

State	Method	Details
MA (MBI, 2011)	• Central office locations were geocoded • 17,800 linear foot service area calculated using road network • 200–foot buffer around network used to define service area • Buffers cropped at town boundaries	Census blocks and road segments that intersected the estimated service area were considered "covered".
IL (PCI, 2011)	• Euclidean buffers of varying sizes were generated around each central office location • Buffers were clipped at wirecenter service area boundaries to estimate coverage areas • Blocks assigned coverage if they were located within modeled coverage areas	The use of Euclidean buffers for the process generates error. See Grubesic (2005) for details.
WV (WVGES, 2011)	• Used DSLAM locations (as points) • Used a 1,000-foot snapping tolerance to assign DSLAM points to a street centerline network. • 15,000 linear foot service area calculated using street centerlines • Service areas were clipped at the wire center service boundaries	These derived coverage areas were used to select covered census blocks and street segments (for large blocks). No details were provided regarding how this geocomputational procedure was executed (e.g., intersection).

for calculating xDSL coverage. For example, the state of Massachusetts (MBI, 2011) uses a 17,800-foot (5.4 km) network distance criterion with an additional buffer of 200 ft. (~61 m) around each street segment. Covered blocks are then selected using an intersection procedure in a geographic information system. On the other hand, West Virginia (WVGES, 2011) uses a 15,000–foot (4.6 km) network distance criterion, but provides no definitive information on how coverage is defined for blocks within this zone (e.g., intersection or completely within). Illinois (PCI, 2011) appears to use a Euclidean-based distance buffer of varying distances to calculate xDSL coverage, but details of associated block coverage estimates are sparse.

Aside from the obvious implications with reporting xDSL provision status for blocks when distance criteria diverge, subsequent geocomputational criteria and methods are also relevant, especially around the margins of estimated xDSL coverages. For example, consider Figure 4.7, which highlights an 18,000-foot network buffer around each CO for Travis County, TX. As one would expect, the xDSL coverage polygons are irregularly shaped, corresponding to local road networks and associated topographical complexities within the region (e.g., lakes and rivers). In this context, the most important question is, how do the designated entities use this type of information to derive xDSL coverage for blocks?

o   Central Office

18,000 ft. Service Area

Travis County

N

Figure 4.7 Derived xDSL coverage areas, Travis County, Texas.

Before we delve into the empirical example, please recall the post-processing rules used by the NTIA (2014b, 1) to estimate the percentage of households covered by broadband service for blocks less than two square miles in size:

> For the purposes of this calculation, when a provider offers service within a block that is less than two square miles, we assume that the provider is offering service throughout the block.

Figure 4.8 provides a drastically simplified snapshot of the vast differences in broadband service coverage estimations when using alternative geocomputational

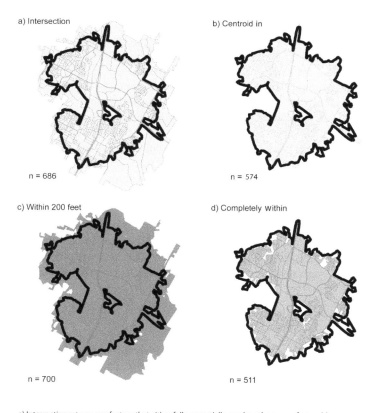

a) Intersection
n = 686

b) Centroid in
n = 574

c) Within 200 feet
n = 700

d) Completely within
n = 511

a) Intersection returns any feature that either fully or partially overlaps the source feature(s)

b) A target feature is selected by this operator if the centroid of its geometry falls into the geometry of the source feature or on its boundaries

c) This operator creates buffers using the buffer distance around the source features and returns all the features intersecting the buffer zones

d) To be selected, all parts of the target features must fall inside the geometry of the source feature(s) and cannot touch the source's boundaries

Figure 4.8 Potential variations in broadband coverage sets.

approaches for blocks *less than 2 square miles in size*. Aside from the important differences in the number of blocks "covered" by using the different spatial selection operators (intersection = 686, completely within = 511, etc.), the spatial characteristics of the selected blocks matter. For example, when using an intersection operator (Figure 4.8a), several larger blocks (in the southeastern portion of the estimated xDSL service area are included in the coverage set. Notice that the bulk of these blocks' geographic area is located outside the 18,000 ft. coverage distance. Thus, it is highly likely that xDSL is not available in the majority of these block regions because the households are located too far away from the CO to receive service.

Figure 4.9 provides the most instructive example of this bias. The maximum xDSL service range is denoted by the black line (18,000 ft.), but the blocks identified as part of the CO coverage set (using an intersection operator) are highlighted by the semi-transparent overlay. Two blocks associated with the tract home development remain *uncovered* (i.e., "orphaned") by the coverage set. This is an accurate assessment, because both blocks are well-beyond the 18,000–foot service zone. However, many of the homes located between these uncovered blocks, and some homes situated even further away from the uncovered blocks are considered within range of xDSL service when using the intersection operator. This is not accurate and reflects the chaotic nature of block group geometries on the margins of this particular xDSL coverage area. Not only are the erroneously selected blocks somewhat larger in geographic extent when compared to their contiguous neighbors, their shape is irregular and they happen to intersect with the 18,000-foot coverage area. This type of error leads to drastic overestimates in xDSL coverage for the region. Again, this is but one example in Travis County, TX. There are hundreds of thousands of examples where this type of loose overestimation is not only possible, but likely. As noted in Table 4.1., the state of Massachusetts uses an intersection-based approach for defining xDSL service areas and it is likely that many other states do this too. Worse, when combined with the NTIA post-processing rules concerning coverage for small blocks (see above), errors abound in the NBM.

Data asymmetries

Finally, there are several caustic data asymmetries within the NBM that, apparently, have gone unnoticed by the broadband community. The first two iterations of the NBM (June 2010 and December 2010) relied on Census 2000 block geometries for data aggregation and reporting. For the June 2011 iteration, the NTIA requested that state-level data collection entities convert to the Census 2010 block geometries (PCI, 2011). This is an important conversion because in 2000, there were 8,262,363 unique Census blocks for the United States and Puerto Rico (Census, 2000). By 2010, this number had grown to 11,155,486 (Census, 2010). In their native format, this means that a straight comparison between 2000 and 2010 Census geometries is *not* possible. A direct comparison between the first two iterations of the NBM and its subsequent releases also appears to be messy.

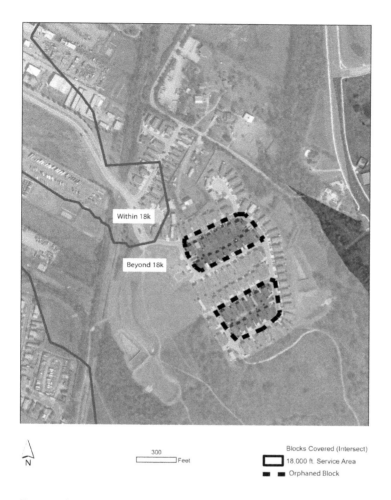

Within 18k

Beyond 18k

300
Feet

N

Blocks Covered (Intersect)
18,000 ft. Service Area
Orphaned Block

Figure 4.9 xDSL coverage errors, Travis County, Texas.

For instance, a more careful inspection of the June 2011 release of the NBM and its associated state grantee methodologies reveals problems. Of the 46 states that issued a report for the June 2011 NBM data, only 29 were definitively using Census 2010 geometries.[13] At a minimum, 14 states appear to be using Census 2000 data for tabulating provider information and estimating broadband availability.[14] Finally, three states (HI, CA, UT) did not provide any information on the vintage of the geographic base files used for reporting broadband.[15] It would appear, therefore, that the June 2011 release of the NBM is an odd mix of Census 2000 and Census 2010 data, but it is nearly impossible to be sure. To reiterate, the exact nature of the data used by many states is difficult to discern, based on uneven quality and detail contained within the grantee methodology reports.

For those states that did use 2010 geometries, the NTIA provided a national "crosswalk" table that was structured to mitigate differences in geometry between Census 2000 and Census 2010, and facilitate an easy conversion between NBM releases. Unfortunately, according to Partnership for a Connected Illinois (PCI, 2011) and many other state grantees, including New Jersey (NJBMP, 2011) and Maine (Sewall, 2011), there were serious problems with the NTIA supplied conversion table. Specifically, these grantees noticed that the NTIA conversion table left "holes" in the converted block tessellations. In other words, the conversion table did not work and some blocks went missing. Regrettably, many other states did not notice these problems.

Consider the case of Illinois. When using the NTIA supplied conversion/ crosswalk table, PCI (2011) identified 605,038 blocks covered by broadband for June 2011, but noticed thousands of geometric gaps in the converted coverages. Again, these gaps corresponded to missing blocks. When using a supplemental overlay process developed internally by PCI to mitigate identified geometric errors caused by the conversion table, Illinois was able to identify an additional 47,564 blocks covered by broadband. This is a massive difference in broadband outcomes and a remarkable problem for the integrity of the NBM. Why? If these types of problems in using the crosswalk table exist for Illinois, they exist elsewhere. Sadly, many states simply accepted the derived results from using the crosswalk table and performed no additional diagnostic and/or quality control measures. It gets worse. Consider what happened in the state of New Jersey (NJBMP, 2011, 10):

> The analysis of the survey data identified some instances where a survey respondent identified their service provider and then the service provider's data did not show coverage in that respondent's Census Block. Further analysis indicated that a number of these instances occurred in "gaps" or "holes" in submitted provider coverage data.

Over 1,400 holes in coverage were identified for a single provider (Comcast; NJBMP, 2011), and hundreds of additional holes were identified for other providers throughout the state. To be clear, these were not the result of the NTIA supplied conversion table. These were problems with the coverages supplied by broadband providers. Regardless, it is not difficult to imagine how these errors begin to comingle (missing blocks, provider coverage holes, etc.) in the NBM, drastically compromising its integrity.

Conclusion

This chapter has covered, no pun intended, a lot of ground concerning available broadband data in the United States. Amazingly, we have only scratched the surface here. This chapter made no attempt to rigorously discuss broadband speed data. This was intentional. Broadband speed data in the United States is highly controversial, difficult to measure and remain a moving target (Grubesic, 2015). Again, given our space limitations, there is no way that a few paragraphs would suffice. We encourage readers to consult recent work by Bauer, Clark, & Lehr, (2010), Canadi, Barford, & Sommers, (2012) and Grubesic (2015), among others to get a better idea of exactly how difficult it is to acquire and summarize variations in broadband speeds for the U.S. In addition, there are many other sources of broadband data that were not discussed, particularly related to infrastructure systems. Although CO locations were used to highlight variations in xDSL coverage for Travis County TX, this chapter did not detail the nuances of wire center service areas, remote digital subscriber line access multiplexer huts (DSLAMs) or the impacts that these infrastructure elements have in determining xDSL service areas within a region (Grubesic & Murray, 2005; Grubesic, 2012a). Fiber routes and fiber lit buildings would also provide a more complete picture of broadband telecommunication infrastructure for a community. Where wireless infrastructure is concerned, understanding gaps in coverage is particularly important for rural and remote areas. Fortunately, there are several global/national repositories of data for cellular towers that can be used for this purpose. Some are open and freely available to analysts (e.g., antennasearch.com, cellmapper.net and opencellid.org), while others need to be purchased (e.g., towermaps.com and geo-tel.com).

As illustrated by this chapter, the etiology of broadband data uncertainties is varied, ranging from incompleteness and divergent methodologies for deriving broadband coverage, to data asymmetries and an inadequate understanding of spatial non-stationarity during the data modeling process.[16] Regardless of the source of data uncertainty, the implications for deepening our understanding of broadband provision, availability, accessibility and equity are substantial, as are the challenges that data uncertainty presents for developing meaningful telecommunications policy.

The limitations of FCC Form 477 data are well known and have been discussed, at length, by many (Grubesic, 2004; Flamm, Friedlander, Horrigan, & Lehr, 2006; Mack & Grubesic, 2014), but these data continue to be unintentionally misused and abused by scholars who have an incomplete understanding of their quirks. In particular, many research efforts mistakenly blend December 2004 (and earlier) data with June 2005 (and later) data collected at the ZIP code level (for example, see Atasoy, 2013; Dettling, 2013; Lelkes, Sood, & Iyengar, 2014). As detailed in this chapter, these versions of the Form 477 data do not measure the same thing.

It is also easy to misunderstand, misinterpret and misapply data from the NBM. Although the NTIA has made a herculean effort to administer the data collection efforts for the U.S., many problems remain. In particular, the empirical results detailed in this chapter suggest that the NBM likely overestimates wireline provider coverage for small blocks throughout the U.S. It is illogical to assume that when a provider offers service within a block that is less than two square miles in size, that the provider is offering service throughout the block. Unfortunately, this is exactly what the NTIA and NBM assume.

There are several ways in which the data uncertainties and errors highlighted in this chapter can be mitigated. First, users of these data need to become more educated regarding collection limitations, data models and processing procedures used by designated entities during the data wrangling process. Although the NTIA has tried to ensure homogeneity, variations remain and this does impact data quality. That said, users bear the ultimate responsibility for any results generated by applications and/or models that use these data. Second, the NTIA should require a more consistent and detailed grantee methodology template to be submitted with each round of the NBM. This should include more information on how broadband service areas are estimated, data quality/confidence metrics, geoprocessing operators used and special section that covers known limitations or problems with the local provider data. For example, some of the state designated entities provided fantastically detailed methodology statements for June 2011 (e.g., Alabama), while others provide almost no meaningful information (e.g., Hawaii). This should be standardized and a consistent, detailed report should be provided by all designated entities. Third, the NTIA needs to develop a suite of more logical and coherent strategies for tabulating and reporting broadband data. As illustrated throughout this chapter, many of the assumptions/rules made by the NTIA are nonsensical. Broadband service availability is not homogeneous in small blocks. The use of a global, 500foot (0.15 km) buffer on roads located in large blocks when estimating broadband coverage statistics is misinformed because offset distances in suburban, rural and remote communities are highly variable. The lack of differentiation between residential and business broadband providers leads to a significant overestimation of competition throughout the U.S. These providers do not serve the same markets.

In sum, elements of data uncertainty permeate many complex databases, including the NBM. When data errors comingle, measurements and models become less robust (Murray & Grubesic, 2012). Ultimately, this negatively impacts policy development and can lead to the mismanagement of scarce resources. Fortunately for the NBM, there is a path forward. Although much of the data uncertainty embedded within the existing iterations of the map (June 2010 – December 2013) can no longer be addressed, it is possible to amend current practices and make subsequent iterations of the map more stable and meaningful. Of course, this would take significant time, effort and resources, but broadband is too important to the U.S. and its growing information economy to ignore this need.

Notes

1 The FCC (2000) considered high-speed lines as connections that delivered services at speeds exceeding 200 kbps in at least one direction. Advanced services lines were capable of 200 kbps in both directions.

2 By definition (FCC, 2014, 1), a facilities-based provider is any entity that; "1) owns the portion of the physical facility that terminates at the end-user premises or obtains the right to use dark fiber or satellite transponder capacity as part of its own network to complete such terminations; 2) it obtains unbundled network element (UNE) loops, special access lines, or other leased facilities that terminate at the end-user premises and provisions/equips them as broadband; 3) it provisions/equips a broadband wireless channel to the end-user premises over licensed or unlicensed spectrum; or 4) it provides terrestrial mobile wireless service using its own network facilities and spectrum for which it holds a license, manages, or has obtained the right to use via a spectrum leasing arrangement."

3 In the late 2000s, there were approximately 32,000 ZIP code areas in the United States (Grubesic, 2008). This did not include ZIP code points, which correspond to high-volume mail generators or receivers (large corporation, university, etc.).

4 There were some exceptions, but these were limited to areas where remote DSLAMs were installed (Grubesic & Horner, 2006). These installations were exceptionally rare in the early 2000s.

5 The exact text from the FCC Form 477 (FCC 2005, 348) report is: "This is a list of geographical zip codes where service providers have reported providing high-speed service to at least one customer as of December 31, 2004. No service provider has reported providing high-speed service in those zip codes not included in this list. An asterisk (*) indicates that there are one to three holding companies reporting service to at least one customer in the zip code. Otherwise, the list contains the number of holding companies reporting high-speed service. The information is from data reported to the FCC in Form 477." The corresponding record for 89049 in the FCC report (http://tinyurl.com/lwslgr5) appears as follows:

State abbreciation	Zip Code	Number of Holding Companies
NV	89046	*
NV	89049	*
NV	89052	6

6 FCC speed tiers are as follows: 1st Generation or greater ($\geq$200 kbps), Tier 1 or greater ($\geq$ 768 kbps), Tier 2 or greater ($\geq$ 1.5 Mbps, Tier 3 or greater ($\geq$ 3 Mbps), Tier 4 or greater ($\geq$ 6 Mbps), Tier 5 or greater ($\geq$ 10 Mbps), Tier 6 or greater ($\geq$ 25 Mbps) and Tier 7 ($\geq$ 100 Mbps).

7 ME (58%), KY (58%), WV (57%), FL (54%), NC (53%), RI (51%), DC (49%), AL (48%), LA (47%), NM (47%), MD (41%), CO (39%), MO (35%), VA (27%).

8 Data portability refers to the ability of data to be used across a variety of interoperable applications. Associated spatial data infrastructures (SDI) are critical to this process and are now widely accepted for exchanging geospatial data amongst organizations (Kiehle, Greve, & Heier, 2007).

9 At the time, Anne Neville (2011), Director, State Broadband Initiative, noted that the top 10–15 providers by broadband technology were included for each state in the NBM, comprising close to 95% of the broadband market. These reported statistics could not be verified independently by the authors.

10 Data sources for verification of broadband provision included digital orthophotos to assist with field validation efforts, and global positioning satellite receivers to verify the locations of central offices, among other things (Connected Texas, 2014)

11 The NTIA (2014b) suggests that the data undergo a rigorous quality review prior to its use in the NBM. This includes cross-checking metadata, making sure the data conform to established geodatabase rules, topology checks, etc.

12 The NTIA also provides an "analyze table" function for the NBM which providers analysts with derived measures of broadband access based on local household and/or population counts. For example, the "prov_gr4" variable represents the percentage of households with access to four or more providers for a block. These data are no less uncertain than the comma delimited or shapefile formats provided by the NTIA. See our discussion in this chapter for more details.

13 AL, AR, CO, CT, DC, ID, IL, KY, LA, MA, MD, ME, MO, NC, NE, NH, NJ, NM, NY, OH, OK, PA, RI, VA, VT, WA, WI, WV, WY.

14 AK, FL, GA, IA, KS, MI, MN, MS, NV, OR, SC, SD, TN, TX.

15 There is a possibility that more definitive information on the Census geometries used by each state was provided on the Grantee Workspace, but this information is not available to the public.

16 Spatial non-stationarity is a term used to describe modeled relationships that are not constant across space. As detailed by Fotheringham, Charlton, & Brunsdon (2012), this means that the process underlying the modeled relationship changes with spatial context.

References

Anselin, L. (1989). What is special about spatial data? Alternative perspectives on spatial data analysis. National Center for Geographic Information and Analysis. Retrieved from http://tinyurl.com/q4nuv6u

Atasoy, H. (2013). The effects of broadband internet expansion on labor market outcomes. *Industrial & Labor Relations Review, 66*(2), 315–345.

Bauer, S., Clark, D.D., & Lehr, W. (2010). *Understanding broadband speed measurements*. Telecommunications Policy Research Conference. Arlington, VA. Retrieved from http://tinyurl.com/nx7hauv

Broadband Data Improvement Act of 2008 (BDIA), Pub. L. No. 110-385 (Oct. 10, 2008).

Canadi, I., Barford, P., & Sommers, J. (2012, November). Revisiting broadband performance. In *Proceedings of the 2012 ACM conference on internet measurement,* (pp. 273–286). ACM.

Connected Nation [Connected Texas]. (2014). *Official April 2014 update submission to the National Telecommunications and Information Administration under the state broadband initiative grant program for the State of Texas*. Retrieved from http://www2.ntia.doc.gov/files/broadband-data/TX-NBM-SHP-Dec-2013.zip

Dettling, L.J. (2013). *Broadband in the labor market: The impact of residential high speed internet on married women's labor force participation*. Retrieved from http://tinyurl.com/nb6ej3a

Federal Communications Commission [FCC]. (2000). *Federal Communications Commission releases data on high-speed services for internet access*. Retrieved from http://tinyurl.com/ls42fc4

Federal Communications Commission [FCC]. (2013). *Modernizing the FCC Form 477 data program*. Retrieved from http://www.fcc.gov/document/modernizing-fcc-form-477-data-program

Federal Communications Commission [FCC]. (2014). *Who must file Form 477?* Retrieved from http://transition.fcc.gov/form477/WhoMustFileForm477.pdf

Flamm, K., Friedlander, A., Horrigan, J., & Lehr, W. (2007). *Measuring broadband: Improving communications policymaking through better data collection.* Pew Internet & American Life Project. Retrieved from http://tinyurl.com/qhzc634

Fotheringham, A.S., Charlton, M., & Brunsdon, C. (1996). The geography of parameter space: An investigation of spatial non-stationarity. *International Journal of Geographical Information Systems, 10*(5), 605–627.

Goldberg, D.W., Wilson, J.P., & Knoblock, C.A. (2007). From text to geographic coordinates: The current state of geocoding. *URISA Journal, 19*(1), 33–46.

Grubesic, T.H. (2003). Inequities in the broadband revolution. *The Annals of Regional Science, 37*(2), 263–289.

Grubesic, T.H. (2004). The geodemographic correlates of broadband access and availability in the United States. *Telematics and Informatics, 21*(4), 335–358.

Grubesic, T.H. (2008). Zip codes and spatial analysis: Problems and prospects. *Socio-Economic Planning Sciences, 42*(2), 129–149.

Grubesic, T.H. (2012a). The US national broadband map: Data limitations and implications. *Telecommunications Policy, 36*(2), 113–126.

Grubesic, T.H. (2012b). The wireless abyss: Deconstructing the US National Broadband Map. *Government Information Quarterly, 29*(4), 532-542.

Grubesic, T.H. (2015). The broadband provision tensor. *Growth and Change.* DOI: 10.1111/grow.12083

Grubesic, T.H., & Horner, M.W. (2006). Deconstructing the divide: extending broadband xDSL services to the periphery. *Environment and Planning B: Planning and Design, 33*(5), 685.

Grubesic, T.H., & Matisziw, T.C. (2006). On the use of ZIP codes and ZIP code tabulation areas (ZCTAs) for the spatial analysis of epidemiological data. *International Journal of Health Geographics, 5*(1), 58.

Grubesic, T.H., & Murray, A.T. (2004). Waiting for broadband: Local competition and the spatial distribution of advanced telecommunication services in the United States. *Growth and Change, 35*(2), 139–165.

Grubesic, T.H., & Murray, A.T. (2005). Geographies of imperfection in telecommunication analysis. *Telecommunications Policy, 29*(1), 69–94.

IDInsight (2011). *Verification analysis of the National Broadband Map.* Retrieved from http://tinyurl.com/3opr85tS.

IN.gov. (2014). *Indiana broadband mapping project.* Retrieved from http://www.in.gov/gis/Broadband.htm

Kiehle, C., Greve, K., & Heier, C. (2007). Requirements for next generation spatial data infrastructures-standardized web based geoprocessing and web service orchestration. *Transactions in GIS, 11*(6), 819–834.

Kruger, L.G. (2009). Broadband infrastructure programs in the American Recovery and Reinvestment Act. Congressional Research Service, Library of Congress. Retrieved from http://fpc.state.gov/documents/organization/122977.pdf

Lennett, B., & Meinrath, S. (2011). Map to nowhere. *Slate.* Retrieved from http://www.slate.com/articles/technology/technology/2011/05/map_to_nowhere.html

Lelkes, Y., Sood, G., & Iyengar, S. (2014). *The hostile audience: the effect of access to broadband internet on partisan affect.* Retrieved from http://www.gsood.com/research/papers/BroadbandPolarization.pdf

Mack, E.A., & Grubesic, T.H. (2009). Forecasting broadband provision. *Information Economics and Policy, 21*(4), 297–311.

Mack, E.A., & Grubesic, T.H. (2014). US broadband policy and the spatio-temporal evolution of broadband markets. *Regional Science Policy & Practice*, *6*(3), 291–308.

Matisziw, T.C., Grubesic, T.H., & Wei, H. (2008). Downscaling spatial structure for the analysis of epidemiological data. *Computers, Environment and Urban Systems*, *32*(1), 81–93.

Massachusetts Broadband Institute [MBI]. (2011). *Methodologies used to create and validate broadband datasets*. Retrieved from http://tinyurl.com/obrlsw7

Meinrath, S. (2011). The FCC needs more fixes, fewer excuses for the National Broadband Map. *Slate*. Retrieved from http://tinyurl.com/c45uh3a

Mrozinski, R.D., & Cromley, R.G. (1999). Singly-and doubly-constrained methods of areal interpolation for vector-based GIS. *Transactions in GIS*, *3*(3), 285–301.

Murray, A.T., & Grubesic, T.H. (2012). Spatial optimization and geographic uncertainty: Implications for sex offender management strategies. In M. Johnson (Ed.), *Community-based operations research* (pp. 121–142). New York: Springer.

Murray, A.T., Grubesic, T.H., Wei, R., & Mack, E.A. (2011). A hybrid geocoding methodology for spatio-temporal data. *Transactions in GIS*, *15*(6), 795–809.

National Telecommunications and Information Administration [NTIA]. (2011a). *National Broadband Map datasets, June 2010*. Retrieved from http://tinyurl.com/nkglklt

National Telecommunications and Information Administration [NTIA]. (2011b). *National Broadband Map: Data comparison methodology*. Retrieved from http://tinyurl.com/3sjv37u

National Telecommunications and Information Administration [NTIA]. (2013a). *Grantee methodologies*. Retrieved from http://tinyurl.com/nn8mccz

National Telecommunications and Information Administration [NTIA]. (2013b). US Dept of Commerce, National Telecommunications and Information Administration, State Broadband Initiative (SHP format December 31, 2013).

National Telecommunications and Information Administration [NTIA]. (2014a). *National Broadband Map datasets*. Retrieved from http://www2.ntia.doc.gov/broadband-data

National Telecommunications and Information Administration [NTIA]. (2014b). *National Broadband Map: Post processing data*. Retrieved from http://tinyurl.com/laft6s9

Neville, A. (2011). Personal communication. September 22, 2011.

New Jersey Broadband Mapping Project [NJBMP]. (2011). *Methodology report on data integration and validation procedures*. Retrieved from http://tinyurl.com/obrlsw7

NYC [NYCBroadbandMap]. (2014). *New York City broadband map*. Retrieved from http://www.nycbbmap.com/

Ookla. (2014). Speedtest. Retrieved from https://www.ookla.com/

Partnership for a Connected Illinois [PCI]. (2011). *Partnership for a Connected Illinois narratives and methodologies*. Retrieved from http://tinyurl.com/obrlsw7

Reibel, M., & Bufalino, M.E. (2005). Street-weighted interpolation techniques for demographic count estimation in incompatible zone systems. *Environment and Planning A*, *37*(1), 127–139.

Sewall. (2011). *Maine SBDD Data Submittal to NTIA Technical Whitepaper*. Retrieved from http://tinyurl.com/obrlsw7

Strover, S. (2001). Rural internet connectivity. *Telecommunications Policy*, *25*(5), 331–347.

United States Census Bureau. (2000a). *Census 2000 data products*. Retrieved from http://tinyurl.com/pcskwcl

United States Census Bureau. (2000b). *Census blocks and block groups*. Retrieved from http://tinyurl.com/psla95k

United States Census Bureau. (2010). *2010 Census tallies of census tracts, block groups & blocks*. Retrieved from http://tinyurl.com/osucpvw

Verizon. (2014). *Verizon wireless coverage locator*. Retrieved from http://tinyurl.com/ox4y6dq

West Virginia Geological and Economic Survey [WVGES]. (2011). State broadband mapping methodology. Retrieved from http://tinyurl.com/obrlsw7

Xie, Y. (1995). The overlaid network algorithms for areal interpolation problem. *Computers, Environment and Urban Systems, 19*(4), 287–306.

Zandbergen, P. A. (2009). Geocoding quality and implications for spatial analysis. *Geography Compass, 3*(2), 647–680.

5   The spatial distribution of broadband provision

With the release of the National Broadband Plan (the *Plan*) in March 2010, the Federal Communications Commission (FCC) outlined a broad and relatively complex strategic agenda for improving broadband infrastructure in the United States (FCC, 2010a). Although broadband continues to be a moving target, it is defined by download (i.e., to the premises) speeds of at least 4 megabytes per second (Mbps) and upload (i.e., from the premises) speeds of at least 1 Mbps for the U.S (FCC, 2010b). In a recent survey from the Pew Research Internet Project (2013), results suggest that 70% of adults in the U.S. (age 18 and older) have a broadband connection at home. Further, in its eighth broadband progress report, the FCC (2012) suggests that 100 million people do not subscribe to the Internet and over 19 million cannot obtain fixed broadband internet services. These results are not surprising when one considers that the United States ranks 16th in the world in fixed broadband subscriptions per 100 residents (29.3), trailing Canada (32.8), France (37) and many others (OECD, 2013).

As noted by Grubesic (2012a), these statistics drastically undersell the complexities associated with both broadband provision and subscription in the U.S. Consider, for example, the 19 million Americans (5.6% of the U.S.) who, for one reason or another, cannot obtain a wireline connection to the Internet. It is important to note that the FCC (2012) statistics say nothing about how many of these residents can obtain a satellite broadband connection, nor how many of them use some type of alternative wireless broadband platform for access (see Chapter 3 for details). Thus, while the federal government, policy analysts and many others continue to voice concerns over the digital divide (Townsend, Sathiaseelan, Fairhurst, & Wallace, 2013; Haffner et al., 2014), gaps in broadband provision and availability may not be as troubling as one might expect. For example, one simply needs to put broadband penetration in context. U.S. telephone subscription rates reached an all-time high of 96% in 2009 (Benton Foundation, 2009).[1] Again, *subscription* is not equivalent to *penetration*, however, considering that 94.4% of the U.S. population can access a fixed broadband line (FCC, 2012), one must ask if the digital divide is still a relevant national policy concern?

Reinforcing the philosophical (and more pragmatic) undercurrents of the digital divide, it is interesting to note that nearly six years after $7.2 billion in federal stimulus funds were allocated to improve broadband provision in the United States (ARRA, 2009), it is no longer clear if the traditional view of the broadband divide remains salient. For example, recent research suggests that the broadband divide is far more complex than the traditional notion of the "haves" and "have-nots" or "core vs. periphery" in the United States (Grubesic, 2015). This body of work suggests the divide has evolved into a multifaceted predicament that includes issues associated with quality of service (Grubesic, Matisziw, & Murray, 2010), adoption (Whitacre & Mills, 2007; Rosston, Savage, & Waldman, 2010) and political participation (Sylvester & McGlynn, 2010). Further, questions concerning the impact of broadband on local and regional economic development (LaRose, Strover, Gregg, & Straubhaar, 2010; Mack et al., 2011) and its growing relevance to cloud based infrastructures (Kloch, Petersen, & Madsen,, 2011) and e-science (Andronico et al., 2011) are beginning to emerge.

Thus, although broadband is widely viewed as an incredibly important technological resource for the United States, a fundamental understanding of access, accessibility, availability, affordability, adoption and use remains obscured, as do the socio-economic and business impacts of broadband. There are three major reasons for this (Grubesic, 2012a). First, as detailed in previous chapters (see Chapter 3), broadband technology is dynamic – both wireline and wireless platforms are constantly evolving. As a result, several generations of telecommunication technologies remain in use throughout the United States. This still includes dial-up, integrated services digital networks (ISDN), cable and xDSL, fiber and the various types of wireless broadband. Second, broadband markets are also dynamic. Competition, or lack of it, impacts price, quality of service (QoS) and many other facets of the broadband provision and consumption. This landscape is also impacted by a complex regulatory environment which involves policies at the national, state and local levels (Flamm & Chaudhuri, 2007). Again, one needs to look no further than the impending $45 billion dollar merger between Comcast and Time Warner. Finally, the data associated with broadband provision, pricing and a multitude of other factors has been of extremely poor quality in the United States during the past decade (Greenstein, 2007; Grubesic 2012a, 2012b). Thus, although the National Broadband Map (NBM) and its data have limitations (see Chapter 4), in many ways the NBM is a significant improvement over the ZIP-code level data provided by the FCC during the late 1990s and early 2000s and subsequent efforts to monitor broadband at the Census tract level.

What is needed now, more than ever, is a comprehensive assessment of the spatial distribution of broadband provision in the United States. Without a viable and accurate snapshot of the broadband landscape, the truly pressing questions concerning linkages between broadband and economic development, QoS and competition, and gaps in broadband availability across different socio-economic and demographic groups will remain difficult to address. Therefore, the purpose of this chapter is to develop a better understanding of the spatial dynamics of broadband availability in the U.S using data from the NBM.

For example, which metropolitan areas in the United States have the highest levels of broadband availability? Are there significant inter-regional differences between metropolitan areas? Do availability gaps between rural and urban areas remain? Again, although the NBM is not perfect, it does serve as the official federal database for documenting broadband availability in the U.S. Thus, this chapter will leverage what we believe to be the most stable release of the NBM data, December 2010, for analyzing broadband provision in the United States.[2] Care will be taken to identify any systematic biases in these data, warning readers when and if these problems might impact interpretation. Further, although this chapter is primarily exploratory in nature, the results of this analysis can provide policymakers and planners with an excellent starting point for launching confirmatory statistical efforts for evaluating key connections between broadband availability and core social and economic development questions facing regions.

The case for broadband regions

One of the core tenets of region development theory is the existence of regional clusters of economic growth, innovation and activity (Bergman & Feser, 1999; Belleflame, Picard, & Thisse, 2000; Gordon & McCann, 2000). For example, Silicon Valley, is often cited as a success story for regional development (Saxenian, 1991, 1996; Kenney, 2000), where entrepreneurialism, tightly knit production networks, a deep pool of skilled labor and an agglomeration of quaternary service-providers (e.g., marketing and law) served to jump-start the preeminent regional innovation machine in the United States (Kenney, 2000). Consider, for example, that there were 15,057 patents registered in Silicon Valley alone during 2012. In 2013, it maintained a 77% share of venture capital investments for California and a 39% share for the entire United States. Silicon Valley had 20 initial public offerings (IPOs) in 2013, which represents a 17% increase over 2012 (SVI, 2014). The reasons that regional and/or industrial clusters develop in places like Silicon Valley are multifaceted, but a key element fueling this type of development is a simple one. Clustering occurs so that industries can accrue the benefits associated with *locational proximity* to similar firms, infrastructure (e.g., power, water and telecommunication), resources for labor and favorable tax structures amongst other things (John & Pouder, 2006). Paralleling the development of regional clusters is the emergence of city-regions, which refer to "a functionally inter-related geographic city area comprising a central, or core, city with a hinterland of smaller urban centers and rural areas, which are socially and economically interdependent" (Charles, Bradley, Chatterton, Coombes, & Gillespie, 1999, 1). Again, Silicon Valley serves as an excellent example of a multi-city-region, where formal political boundaries of cities are not good delimiters of the social, cultural or economic interconnections of a place (Friedman & Miller, 1965; Green et al., 2007). Rather, proximal cities often function as an economic system, maximizing the flow of information, ideas and inputs between entities.

As detailed by Grubesic (2006, 2008), a key enabler of the intra- and inter-regional interconnection of economic clusters and city-regions is information and communications technologies (ICTs). Much like the railroad system in the late 1800s and the interstate highway system of the mid-to-late 1900s in the U.S., telecommunication systems have a significant impact on the economic geography of places and their interconnections (Graham & Marvin, 1996; Graham, 1999; Wheeler, Aoyama, & Warf, 2000; Forman, Goldfarb, & Greenstein, 2005; Steinfield, Scupola, & López-Nicolás, 2010; Mack, 2014a; Mack & Rey, 2014). One of the main reasons that industries cluster in a region is so that organizations can transfer complex information between each other without being forced to overcome large distances (Anselin, Varga, & Acs, 1997). However, with the advent of advanced telecommunications technologies, such as broadband, it is increasingly clear that although distance is not dead (Cairncross, 2001), the effects of distance can wane under the right conditions (Johnson, Siripong, & Brown, 2006). For example, the average distance between patent collaborators increased 2.5% annually between 1975 and 1999 (ibid.). Similarly, the work of Petruzelli (2011) suggests that proximity may not be an important factor in successful academic and industrial collaborations. In particular, the most successful university/industry collaborations tend to occur between distant partners (Petruzelli, 2011). Although one cannot directly attribute these outcomes to ICTs and/or broadband, it is clear that ICTs will continue to play a major role in regional development efforts and collaboration within and between city-regions.

Given the strong ties between ICTs and regional development, the concept of broadband regions is a natural one. As detailed in previous chapters, the spatial landscape of broadband is multifaceted and prone to distributional biases of specific technologies and broadband platforms. For example, xDSL technologies are quite limited in service distance from the central office. Most providers try to limit loop lengths (i.e., distance from end-user premises to the central office) to minimize attenuation and ensure service quality. Although remote switches and hybrid systems have extended the reach of digital subscriber lines (Grubesic & Horner, 2006; Grubesic et al., 2010), it is clear that spatial scale, technology and a complex mesh of supply-side and demand-side determinants impact the availability of broadband (Prieger & Lee, 2006; Flamm & Chaudhuri, 2007; Mack, 2014a; Mack & Grubesic, 2014) and that no two regions will have the exact same footprint of broadband services available, nor will regions demand and/or consume broadband in the exact same way.

The result of these different supply- and demand-side landscapes profiles are heterogeneous broadband regions (Grubesic, 2006, 2008), where some locales benefit from a robust set of providers and technologies, while others struggle to achieve any diversity in their local broadband offerings. In the next section, we outline the data and methods for developing a typological profile of broadband provision in the United States that will help deepen our understanding of the patchwork nature provision at the local and regional levels.

Data and methods

Broadband data

NBM data for December 2010 were obtained from the National Telecommunications and Information Administration (NTIA) from their website (http://www2.ntia. doc.gov/broadband-data). As detailed in Chapter 4, the native spatial aggregation of these data is the Census block level. For the purposes of this chapter, block level provision data were aggregated to the block group level.[3] A total of 207,507 block groups is used for analysis. There are several reasons for pursuing an analysis at the block group level. In addition to helping ease visualization efforts, the use of data aggregated to block groups helps smooth error in the database. As noted in Chapter 4, the NTIA collected wireline broadband data in two different ways. For blocks that were less than two square miles in size, local providers reported block numbers where broadband services were provisioned. This is a simple tabulation effort and the total count of unique providers is easily obtained for these units. For blocks that are two square miles in size or greater, providers denoted addresses or road segment data where broadband service was provisioned. This type of tabulation is less user-friendly because there are many duplicate entries for each large block. For example, there is a block in the state of Ohio where AT&T reports 55 unique service addresses. This is not a problem, but it is important to remember that if AT&T is the only provider that serves this block, then one provider is present, not fifty-five. In short, we use both sets of data for this analysis, taking care to aggregate provider counts carefully to the block group level and eliminate any duplicates that may be found in large blocks.

Administrative units

Several supplementary databases are also used for analysis. Core-based statistical areas (CBSA), which are defined as one or more adjacent counties or county equivalents that have at least one core urban area of 10,000 residents or more, plus adjacent territory that has a high degree of social and economic integration with the core (OMB, 2010). This is typically measured by commuting ties (ibid.). There are currently 929 CBSAs in the United States. Within the CBSA database, there are two sub-classifications: 1) Metropolitan Statistical Areas (MSAs), which are urbanized areas of 50,000 or more population and 2) Micropolitan Statistical Areas (μSAs), which correspond to urban clusters of at least 10,000 residents but less than 50,000. As of 2010, there are 388 MSAs in the U.S. and 541 μSAs. Again, it is important to reiterate that both MSAs and μSAs are county-level composites.

Urbanized areas are also used for this analysis. The Census Bureau (2010) identifies two different types of urban areas: 1) Urbanized Areas (UAs) correspond to areas of 50,000 or more residents and 2) Urban Clusters (UCs) represent regions with at least 2,500 residents, but less than 50,000. Rural areas, then, encompass all population and territory not included within an urban area (Federal Register, 2011). To clarify, UAs and UCs form the urban cores of metropolitan and micropolitan statistical areas, respectively (Census, 2010). Each MSA contains a UA of 50,000 residents or more. Each µSA contains at least one UC of at least 10,000 residents, but less than 50,000.[4]

Methodology

Previous work dealing with the spatial distribution of broadband in the United States applied statistical techniques to generate a spatial taxonomy of broadband regions (Grubesic, 2006, 2008). This chapter utilizes a similar approach and employs local indicators of spatial association, namely the local Moran's I statistic (Anselin, 1995), to identify a patterns in broadband provision. The local Moran's I is specified as follows:

$$I_i = z_i \sum_j w_{ij} z_j,$$

(6.1)

Where x_i and x_j are the observations for locations i and j (with mean µ), $z_i = (x_i - \mu)$, $z_j = (x_j - \mu)$, and w_{ij} is the spatial weights matrix with values of 0 or 1. The interpretation of this statistic is relatively straightforward. The local Moran's I will be greater than zero when there is an association of similar values to the location, i, where the index is measured. If the local statistic is less than zero, spatial autocorrelation is also present, suggesting non-similar levels of broadband provision surrounding location i. If there is no spatial association between units, the local Moran's I will have a value of zero. Ultimately, this process generates spatial clusters which are structured as contiguous locations (if one is using a contiguous weight matrix) for which the local Moran's I statistic is significant. For the purposes of this analysis, a multitude of spatial weighting schemes were tested, but ultimately the use of a k-nearest neighbor (k-NN) weights matrix was selected for delineating the broadband regions. There are two reasons for this. First, the overarching goal of developing this typology is to identify locally significant clusters of block groups with similar levels of broadband provision (high, low and in-between). With k-NN, this is easily accomplished and it provides the opportunity for testing the sensitivity of the results across a range of neighborhood structures by adjusting the value of k. Second, block groups are not contiguous throughout the United States. For example, block groups in Key West, FL (an island and the southern-most point in the continental United States) do not share borders with any adjacent land mass. As a result, the use of a standard contiguity measure (e.g., rook or queen) would bias the statistic.[5]

For this analysis, a range of k-NN values were tested, but ultimately a value of 6 was selected for generating broadband regions because it provided the best balance of meaningful, local spatial context without extending the search too far afield and incorporating spurious neighboring units and/or geographic outliers. Table 5.1 provides a simple translation of the spatial taxonomy, mirroring the basic groupings developed by Grubesic (2006). Again, it is important to emphasize here that although the "core vs. periphery" framework is somewhat naïve (Grubesic, 2015), especially given the other important facets of broadband provision (e.g., pricing, speed and QoS), it still provides an excellent foundation for exploring the spatial distribution of broadband provision, regional differences and inequities.

A second facet of this exploratory analysis is the application of the broadband deployment index (BDI) (Grubesic, 2008). The BDI is specified as follows:

$$BDI_i = \left(\kappa_i\left((\alpha_i + \beta_i) - (\varphi_i + \lambda_i)\right)\right) \times 100 \tag{6.2}$$

where i is the index of metropolitan or micropolitan areas,[6] α_i is the percent of block groups classified as *broadband core* in the ith area, β_i is the percent of block groups classified as *islands of availability* in the ith area, φ_i is the percent of block groups classified as *broadband periphery* in the ith area, λ_i is the percent of block groups classified as *islands of inequity* in the ith area, and κ_i is the scalar weight of block groups classified in the ith area as a percentage of all block groups in the MSAs or μSAs. For convenience and interpretive purposes, the BDI is bounded between -100 and 100. As a standardized measure, it allows for the control of market size and provides an effective measure for ranking broadband markets using the spatial taxonomy generated by the LISA statistic. The interpretation of the BDI is relatively straightforward. Block groups that are classified as the broadband core or islands of availability provide a positive contribution to the index, while block groups that are classified as broadband periphery or islands of inequity provide a negative contribution to the index. When BDI values trend toward 100, the MSA or μSA has a stronger broadband provision profile and when BDI values trend toward -100, the MSA or μSA has a weaker palette of broadband provision.

Results

Table 5.1 displays the descriptive statistics for broadband provision in December 2010 at the block group level for a variety of different geographical aggregations. At the national level, the maximum number of providers found in a single block group is 16, located in Baltimore, MD. The most relevant statistic in this table is the average number of providers, by block group, for the entire U.S., which is 2.836. At the very least, this suggests there is some *choice* in broadband provision, on average, for many regions. However, it is also clear from Table 5.1 that the statistical averages associated with broadband provision strongly favor UAs (3.092) and MSAs (2.969) when compared to their smaller urbanized clusters (2.342) and μSA (2.399) counterparts. These results are not unexpected, as the United States has always displayed a strong urban bias in advanced telecommunications infrastructure (Grubesic & Murray, 2002, 2004; Greenstein & Prince, 2006).[7]

Figure 5.1 is a choropleth map that displays counts of broadband providers, by block group, for the United States in December 2010. Figure 5.2 is a more easily interpretable map that highlights the spatial density of broadband providers at the block group level for the United States in December 2010. There are

Table 5.1 A spatial taxonomy of broadband regions

Broadband Core	Block groups displaying high levels of broadband provision that are surrounded by neighboring block groups with similar values.
Broadband Periphery	Block groups displaying low levels of broadband provision that that are surrounded by neighboring block groups with similar values.
Islands of Inequity	Block groups displaying low levels of broadband provision that are surrounded by neighboring block groups displaying relatively high values.
Islands of Availability	Block groups displaying high levels of broadband provision that are surrounded by neighboring block groups displaying relatively low values.
Not significant*	Block groups that are not statistically significant.

*$p = 0.05\%$

a number of patterns worth noting here. First, it is clear that the bulk of high-density areas are located in major urban corridors. For example, Megalopolis, the urban region extending from Boston, MA to Washington, DC is dense with broadband provision. As detailed in previous chapters, large urban areas generally provide the greatest return on investment for broadband providers (Grubesic, 2003). This is reflected by provision density for Megalopolis and many other

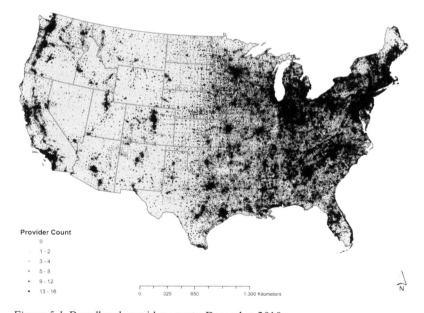

Provider Count
0
1 - 2
3 - 4
5 - 8
9 - 12
13 - 16

0   325   650   1,300 Kilometers

N

Figure 5.1 Broadband provider counts, December 2010.

Provider Count
Density
0.000
0.001 - 0.356
0.357 - 1.326
1.327 - 2.938
2.939 - 5.501
5.502 - 11.403
11.404 - 21110.245

0   325   650   1.300 Kilometers

Figure 5.2 Broadband provider density, December 2010.

major metropolitan regions including Southern California, the San Francisco Bay Area, and Denver and the Front Range, where provision densities are also quite high. However, simplistic measures of provision density can be misleading. The geographic size of block groups factors into this. Block groups tend to be much smaller in urban areas and even those with a handful of providers appear to be dense with options. If the same number of providers exists in rural areas with larger-sized block groups, provision does not appear to be as dense. In an effort to mitigate this uncertainty, additional statistical measures and associated visualizations are needed.

Table 5.2 displays the results of the local Moran's I statistic for broadband provider counts by block group in the United States for December 2010. As detailed previously, these results are generated by using a k-nearest neighbor weight of 6 to derive the associated clusters and/or typology. In this case, $I = 0.7293$ ($z = 623.16$, $p = 0.001$), suggesting a strong level of positive spatial autocorrelation in broadband provider counts by block group for the U.S. Table 5.2 also details the variations in provider averages across each of the regions. As one would expect, core regions display the highest average (4.49) and peripheral regions the lowest (1.79). Given these results, readers may question how any block groups in the broadband core could only have one provider. Similarly, how could any block groups classified in the broadband periphery have four providers? Recall that the statistic used to derive this typology uses spatial weights

Table 5.2 Descriptive statistics for broadband provision by block group, December 2010

United States

n	207,507
min	0
max	16
mean	2.836
SD	1.21

Metropolitan statistical areas		*Micropolitan statistical areas*	
n	168,020	n	21,674
min	0	min	0
max	16	max	8
mean	2.969	mean	2.399
SD	1.20	SD	1.01

Urbanized areas		*Urbanized clusters*	
n	134,120	n	14,597
min	0	min	0
max	16	max	8
mean	3.092	mean	2.342
SD	1.2	SD	0.87

to define local neighborhoods. So, if there is a block group with one provider located within an extremely poor broadband region, it can qualify as a core area. The inverse is also true. Block groups with four providers that are classified as broadband periphery are likely located in regions where their neighboring block groups have significantly more providers (e.g. ten or more). This is a desired feature of the local statistic because it can account for these regional heterogeneities in an intuitive way. We both *want* and *expect* local variation in the provider counts for each derived region because each area and its associated context is different across the United States.

Consider Figure 5.3, which highlights the broadband core regions in the United States. The bulk of the broadband core is found in major metropolitan areas (New York City, Philadelphia, Washington, DC, Seattle, WA, etc.). However, what some readers may find surprising is the strong presence of broadband core areas in states like Indiana and Mississippi. We shared some initial skepticism with these results and immediately revisited the details on how these data were collected and tabulated by independent state agencies. However, where Indiana and Mississippi are concerned, there is a convincing case to be made that the core regions highlighted are offering potential subscribers a wide range of provision choice. We will use examples from Indiana to illustrate how such patterns manifest.

Consider Fishers, IN, an affluent suburban community located north of Indianapolis and part of the broadband core. Most residential locations in this

Figure 5.3 Broadband core, December 2010.

city were reported as being served by at least seven providers in December 2010. For example, one representative address is 12043 Bird Key Blvd, Fishers, IN 46037. This location is served by AT&T Indiana, Comcast, Zayo, On-Ramp Indiana, MegaPath, Covad and Indiana Fiber Network (IFN). This is a lot of choice for one address, certainly above the national average for broadband core regions (4.49), but there are some important caveats with these results worth mentioning. First, companies like IFN and Zayo are not direct providers of residential broadband service. For example, IFN operates a large metropolitan fiber ring in the Indianapolis region and is geared toward providing broadband services to businesses or other large organizations. Second, Covad and MegaPath are the same company, having merged in September 2010 (Megapath, 2010). This merger was too recent to be reflected in the December 2010 NBM database, but it is interesting to note that both carriers are still reported as separate options for subscribers throughout Indiana (and elsewhere) in July 2014 on the Indiana Broadband Map portal (http://tinyurl.com/mz4l3u8). This is not the case for the most recent NBM data (December, 2013), where IFN, Zayo, Comcast, AT&T, Platinum Equity and On-Ramp Indiana are reported as providing service to the Bird Key address.[8] In all, the broadband core regions highlighted in Indiana and Mississippi are strongly representative of the core itself. Indiana averages 4.64 providers within the core, which is slightly above the national average (4.49). The same can be said for Mississippi (4.81). So, while provision in these regions cannot compare with the highest-flying block groups

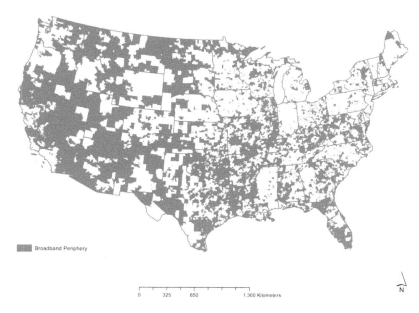

Broadband Periphery

0   325   650   1,300 Kilometers

N

Figure 5.4 Broadband periphery, December 2010.

that are members of the core (e.g. 12–16 providers), there is certainly a lot of choice for subscribers in these areas.

Figure 5.4 displays the broadband periphery for the U.S. in December 2010. Although these regions are considered the least robust areas in terms of broadband availability and provider choice, it is important for readers to remember that broadband *is* available for the bulk of the block groups in this category. For example, nationally, the average number of providers in peripheral regions is 1.793. This is very different from the regional average for the broadband core, but broadband is available nonetheless. In fact, of the 59,750 block groups within the periphery, only 1,337 (2.23%) are without any type of wireline broadband service. The vast majority of these locations correspond to large tracts of wilderness in the Western U.S., remote parts of Appalachia in the Eastern U.S. or extremely rural portions of the Plains, Midwest or South. So, although the geographic coverage of the broadband periphery is vast, these block groups do not represent a "black hole" for broadband service. In most cases, at least one wireline provider is present.

Figure 5.5 illustrates the islands of availability for December 2010 in the United States. With a national average of 3.12 providers per block group, these areas are typically nested within geographic locations strongly associated with the broadband periphery. Consider the island of availability just northwest of Fresno, CA near Millerton Lake (a large recreational destination for the Central Valley). The community of Friant, CA only has a population of 549, but it functions as

Figure 5.5 Broadband islands of availability, December 2010.

a service center for the large recreational area around the lake, casino, marina/ resort, golf club, etc. Thus, although its immediate surroundings are rural/remote, there is enough demand in this area to spur some competition amongst providers. In 2010, three providers were present in Friant, CA, but there were many areas in and around Friant (foothills of the Sierra Nevada mountain range), where no broadband provision is available. This is a fairly representative case for the islands of availability region – isolated pockets of availability in larger regions with limited or no broadband options.

Similar to how the islands of availability are loosely proximal to the broadband periphery, islands of inequity are often proximal to the broadband core. Again, these are areas that have limited broadband options near regions with a much stronger profile of broadband availability. Figure 5.6 displays the islands of inequity for the United States in December 2010. Of the four derived regions, islands of inequity has the smallest number of block group members ($n = 530$), but its average provider count (1.903) is somewhat higher than the broadband periphery (Table 5.3). Once again, the bulk of this region *has* broadband, although provider choice tends to be lower in these block groups than its local neighbors. For example, one can easily revisit the earlier example of the broadband core regions located in Indiana to better understand the spatial dynamics of the islands of inequity. Northwest of Indianapolis are two commuter suburbs that fit this profile, Whitestown and Pittsboro, IN. Both communities had two broadband providers in December 2010; TDS, the seventh largest local exchange telephone company in

Figure 5.6 Broadband islands of inequity, December 2010.

the U.S. (xDSL service) and Brighthouse Networks, the sixth largest cable internet provider in the U.S. The presence of a duopoly within these two communities region is not surprising. Brownsboro, IN, which is only 4 km away from Pittsboro on Highway 136 is a member of the broadband core and has five providers, including Comcast, Bright House, AT&T, Platinum Equity and On-Ramp Indiana. This is a huge difference in provision between two neighboring communities, but one

Table 5.3 Local spatial autocorrelation of broadband providers, December 2010

Local Moran's I	0.7293	
z	623.16	
p	0.001	
Broadband core		*Broadband periphery*
Block Group Count	36,788	59,750
(min) [max]	(1) [16]	(0) [4]
Average	4.4974	1.793
Islands of availability		*Islands of inequity*
Block group count	4,259	530
(min) [max]	(1) [9]	(0) [5]
Average	3.121	1.903

that is illustrative of steep spatial gradients between core regions and islands of inequity. Distance really does matter for broadband provision and this example is certainly illustrative of its importance.

Ranking the regions

Now that a reasonable understanding of the spatial distribution of broadband regions and their associated provision characteristics has been established, it is possible to evaluate metropolitan and micropolitan regions based on the BDI. As detailed previously, there are several core questions concerning broadband provision in the United States that need to be addressed. First, which metropolitan areas in the United States have the highest levels of broadband availability? Again, not only is broadband is a key component of regional competitiveness (Mack, 2014b), it also serves as one of the technological platforms that enables intra- and inter-regional interconnection of economic clusters and city-regions. A second important question to address is determining whether or not gaps in broadband availability remain between rural and urban areas. This facet of a larger literature on the digital divide (Hoffman & Novak, 1998; Compaine, 2001; Van Dijk & Hackler, 2003; Livingstone & Helsper, 2007; Goldfarb & Prince, 2008) remains relevant because broadband is a technology that can, at least in part, mitigate elements of the "rural penalty" that often exists for geographically remote communities. This includes expanding retail shopping options, making social services more efficient and helping diversify local economic activities. Finally, in addition to benchmarking variations in broadband deployment between regions, it is equally important to benchmark variations *within* regions. This is often overlooked by analysts because of the technical challenges associated with capturing and processing disaggregate spatial data, but it is perhaps the best way to motivate local public policy efforts to enhance broadband availability if availability gaps within the larger metropolitan area exist.

Metropolitan areas (MSAs)

Figure 5.7 displays all 377 metropolitan areas in the United States ranked by the derived BDI score for each region. The choropleth shading used for the thematic representation in Figure 5.7 is structured intuitively: light gray and white colors suggest low levels of broadband deployment, darker gray and black colors suggest high levels of broadband deployment. In effect, it is designed to represent the full spectrum of the BDI. The most noticeable patterns correspond to locations where the BDI is highest and lowest for the U.S. For example, portions of the Northwest (Seattle and Portland) and the Colorado Front Range (Denver and Fort Collins) display relatively high BDI values, while places like Great Falls, MT, Fort Myers, FL and El Paso, TX display extremely low BDI values. The average BDI score for all MSAs in the U.S. is -12.07 with a standard deviation of 32.59. This suggests a relatively large spread of values within the BDI, with most MSAs gently trending toward the negative. Again, the negative

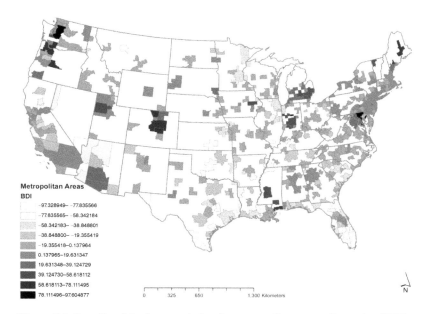

Figure 5.7 Broadband deployment index for metropolitan areas, December 2010.

average value for the BDI is not cause for alarm; it simply captures the relative level of deployment amongst MSAs. Even with a negative BDI score, broadband is still available.

Table 5.4 highlights the top 25 "leaders" and bottom 25 "laggards" for the BDI in the U.S. The leaders represent an interesting mix of both larger and smaller MSAs, with Kokomo, Indiana taking the top spot. Kokomo is an interesting city. It is heavily invested in industrial production (e.g., automobile parts), but is subject to disruptions in employment and overall economic health as industrial production fluctuates (Parkison, 2011). Over the past several years, Kokomo has rebounded economically, ranking 61st amongst all MSAs in gross domestic product (GDP) growth (Kokomo Perspective, 2013; Smith, 2014). With this resurgence of manufacturing and employment in Kokomo, it is likely that the growing pool of middle-class workers in the region are demanding broadband services and spurring competition in the area. This is also complemented by Kokomo's strong profile in higher education. Both Indiana University and Purdue University maintain branch campuses in Kokomo, as does Ivy Tech Community College and Indiana Wesleyan University.

The Baltimore–Columbia–Towson, Maryland MSA ranks second, with a BDI score of 86.58. As mentioned earlier, Baltimore is home to the block group with the single largest presence of wireline broadband providers in the country ($n = 16$). It is also important to remember that the Baltimore MSA is the fourth largest in the United States with a population of 2.7 million, and portions of it serve as bedroom/commuter communities for the Washington, DC area. It has the fourth highest median household income in the United States at nearly $66,654, trailing only Washington ($88,505) San Francisco ($73,563) and Boston ($70,699). In short, demand for broadband in the Baltimore MSA is high and the BDI value for the region reflects this.

Where the laggards are concerned, the Carbondale–Marion, Illinois MSA displays the lowest BDI score (−97.32). Located in Southwestern Illinois, the Carbondale–Marion MSA is relatively small in terms of population (126,745), ranking 311th of all MSAs for 2010 and shows virtually no signs of growth (Census, 2010). The region is largely agrarian, with the major agricultural output consisting of corn and soybeans. Some light manufacturing is also present within the region, including Aisin Seiki Co., Ltd (manufacturer of automobile parts and small electronics). The largest economic and cultural hub in this MSA is Southern Illinois University-Carbondale (SIUC). With an enrollment of over 17,000 students, SIUC is the flagship campus of the Southern Illinois University system. In recent work, Oyana (2011) details the challenges associated with both broadband demand and delivery in this region. In short, because the population density is so low, the broadband providers that do offer service tend to cluster in and around the cities of Carbondale, Marion, Herrin and Murphysboro. The areas surrounding these cities include huge tracts of the Shawnee National Forest that have no broadband provision. Again, this suggests a very steep gradient in broadband deployment for the region, similar to many of the other communities in the U.S. that are located in rural and remote regions.

The Naples–Immokalee–Marco Island, Florida MSA, which is located in Collier County, also displays a very low BDI score (−89.81). Unlike its peer region (Carbondale–Marion), the population is growing rapidly in Collier County. For example, between 2000 and 2010, the population in this region increased by 27.9%, from 251,377 to 321,520. This is after several decades of growth exceeding 75%. However, the bulk of the population in this area is older (similar to the state of Florida), with nearly 50% of the county aged 45 or more. As detailed in previous work (Grubesic, 2003, 2006, 2010; Whitacre, 2010), older populations and broadband deployment tend to be negatively related (this is discussed in more detail in Chapter 9). Thus, it is not surprising to see a lower BDI for major retirement destinations like Naples and Marco Island. In fact, most of the laggard metropolitan areas listed in Table 5.4 (located in Florida, Texas and Arizona) correspond to regions that attract retirees and/or an older demographic.

Urbanized areas (UAs)

Many of the same general trends (and community rankings) uncovered in the BDI calculations for MSAs also exist for the UAs (Table 5.5). The average BDI score for all UAs in the U.S. is −8.89 with a standard deviation of 39.62. Once

Table 5.4 Broadband deployment index: MSA leaders and laggards, December 2010

Rank	MSA	BDI	Rank	MSA	BDI
1	Kokomo, IN	97.60	24	Salt Lake City, UT	39.20
2	Baltimore–Columbia–Towson, MD	86.59	25	Medford, OR	38.73
			377	Carbondale–Marion, IL	−97.33
3	Seattle–Tacoma–Bellevue, WA	79.08	376	Naples–Immokalee–Marco Island, FL	−89.82
4	Colorado Springs, CO	77.39	375	Victoria, TX	−88.31
5	Gulfport–Biloxi–Pascagoula, MS	76.48	374	Sebastian–Vero Beach, FL	−88.09
6	Denver–Aurora–Lakewood, CO	75.61	373	Brownsville–Harlingen, TX	−84.01
			372	Great Falls, MT	−83.71
7	California–Lexington Park, MD	67.82	371	Cape Coral–Fort Myers, FL	−83.59
			370	El Paso, TX	−82.64
8	Bangor, ME	65.24	369	McAllen–Edinburg–Mission, TX	−81.70
9	Bend–Redmond, OR	64.72			
10	Fort Collins, CO	63.02	368	Sebring, FL	−78.95
11	Portland–Vancouver–Hillsboro, OR-WA	59.99	367	Jacksonville, FL	−78.44
			366	Las Vegas–Henderson–Paradise, NV	−78.38
12	Lewiston–Auburn, ME	59.93			
13	Burlington–South Burlington, VT	57.13	365	El Centro, CA	−78.14
			364	Yuma, AZ	−77.37
14	Jackson, MS	56.77	363	Beaumont–Port Arthur, TX	−76.82
15	Hattiesburg, MS	53.29	362	Salinas, CA	−76.24
16	Indianapolis–Carmel–Anderson, IN	51.34	361	Lake Havasu City–Kingman, AZ	−75.98
17	Bloomington, IN	49.98	360	Yuba City, CA	−75.32
18	Olympia-Tumwater, WA	45.01	359	Pine Bluff, AR	−74.98
			358	Miami–Fort Lauderdale–West Palm Beach, FL	−74.31
19	Ann Arbor, MI	44.58			
20	Lafayette–West Lafayette, IN	43.81	357	Reno, NV	−72.43
			356	Baton Rouge, LA	−72.25
21	Lansing–East Lansing, MI	42.58	355	Port St. Lucie, FL	−71.96
			354	Owensboro, KY	−70.83
22	Cedar Rapids, IA	42.42	353	Chico, CA	−69.00
23	Detroit–Warren–Dearborn, MI	40.44			

again, this suggests a relatively large spread of values within the BDI, with most UAs gently trending toward the negative. Where the leaders are concerned, Kokomo remains at the top of the list; but curiously, it is joined by the UAs of Hattiesburg, MS and Cedar Rapids, IA. Although both of these MSAs were in the top 25 nationally, their rankings were 15th and 22nd, respectively. Further, virtually all of the leaders that appeared in the MSA list (Table 5.4) have higher

Table 5.5 Broadband deployment index: UA leaders and laggards, December 2010

Rank	MSA	BDI	Rank	MSA	BDI
1	Kokomo, IN	100	377	Carbondale, IL	−100
2	Hattiesburg, MS	100	376	Dekalb, IL	−100
3	Bangor, ME	100	375	Titusville, FL	−95.29
4	Cedar Rapids, IA	100	374	St. Augustine, FL	−95.29
5	Gulfport-Biloxi-	98.85	373	Harlingen, TX	−93.5
	Pascagoula, MS		372	Sebastian-Vero Beach, FL	−90.57
6	Baltimore, MD	98.79	371	Baton Rouge, LA	−90.37
7	Burlington, VT	97.45	370	Seaside-Monterey, CA	−89.75
8	Westminster-	96.63	369	Bonita Springs, FL	−88.5
	Eldersburgh, MD		368	Great Falls, MT	−87.86
9	Colorado Springs, CO	96.22	367	Jacksonville, FL	−87.19
10	Bend, OR	95.78	366	Lake Havasu City, AZ	−87.11
11	Waldorf, MD	94.03	365	Pine Bluff, AR	−86.01
12	Jackson, MS	88.7	364	Beaumont, TX	−84.56
13	Seattle, WA	87.99	363	Laredo, TX	−84.54
14	Fort Collins, CO	87.82	362	Florence, SC	−84.33
15	Portland, OR	85.24	361	Cape Coral, FL	−84.02
16	Lewiston, ME	84.92	360	Yuba City, CA	−83.25
17	Denver-Aurora, CO	84.8	359	Brownsville, TX	−83.13
18	Pascagoula, MS	84.52	358	El Paso, TX	−83.01
19	Olympia-Lacy, WA	83.87	357	Victoria, TX	−82.33
20	Greeley, CO	82.8	356	McAllen, TX	−80.99
21	Lansing, MI	78.81	355	Lawrence, KS	−80.96
22	Sioux Falls, SD	77.04	354	Wichita, KS	−80.93
23	Portland, ME	75.67	353	Owensboro, KY	−80.69
24	Anderson, IN	74.97			
25	Medford, OR	74.04			

BDI scores in the UA list. For example, the Baltimore–Columbia–Towson, MD MSA had a BDI score of 86.59, but the Baltimore UA has a score of 98.79, a difference of over 12 points. This suggests that the broadband deployment gradient is present within major metropolitan areas and is not simply a trend for exurban, rural or remote locations.

Consider Figure 5.8 which illustrates this broadband deployment gradient for the Baltimore region. In this figure, a kernel density function (Langford & Unwin, 1994; Kelsall & Diggle, 1995) was used to capture the inter-MSA differences in broadband deployment at the block group level. All 1,893 block groups located within the Baltimore–Columbia–Towson, MD MSA were used for analysis and multiple density surfaces were generated using a variety of search radii (5,280 ft. [1,608 m], 10,560 ft. [3,218 m], 15,840 ft. [4,828 m], etc.). To facilitate discussion, a generally representative search distance, 10,560 ft. (two miles; 3.2 km), is illustrated. To generate this surface, each block group centroid

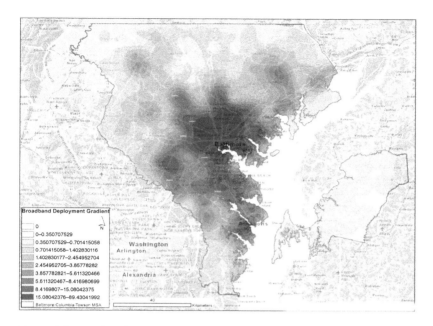

Figure 5.8 Broadband deployment gradient, Baltimore, MD.

is weighted by the number of broadband providers present. Block groups with a high number of providers that are geographically proximal represent areas with a higher deployment density. Those with fewer providers that are not geographically proximal represent areas with a lower deployment density. Of interest here is how the deployment gradient decays as one moves away from central Baltimore and toward the periphery of the Baltimore–Columbia–Towson MSA. One can clearly see the higher density of provision within the core of the MSA (corresponding to the Baltimore UA). Additional areas of elevated provision density also exist near the cities of Columbia (southwest of Baltimore), where there are between 4 and 14 providers in the local block groups, Bel Air (northeast of Baltimore), where there are between 4 and 8 providers in most block groups and Annapolis (southeast of Baltimore), where there are between 5 and 13 providers locally.

This is a significant finding for several reasons. First, it suggests that the geographic distribution of broadband provision is significantly more complex at the local level than once thought. Heterogeneity *within* MSAs is extensive. Thus, it is likely that most analytical efforts attempting to model and/or explain broadband provision, adoption or competition at the local level, need to be using these high-resolution data (e.g., Census tract or smaller). Efforts at the city, county or MSA level do not suffice, especially when exploring questions concerning the digital divide and the impacts of broadband on regional development. Second, the broadband deployment gradient is likely dynamic. Although Baltimore is used as the example in this chapter, the deployment gradient for places like Springfield, Missouri or Boise, Idaho would undoubtedly be different. Clearly, more work is needed in this domain to capture and explain how (and if) broadband deployment decays in a systematic manner from major urban cores. This is particularly important for economic development entities in suburban, exurban and rural locations as they inventory assets and market their communities to businesses. Although research has suggested broadband is not a panacea for all community issues (Hales, Gieseke, & Vargas-Chanes, 2000), it is certainly an increasingly important part of the critical infrastructure that impacts business location decisions.

Micropolitan areas (μSAs)

The final group of locations compared in this analysis is micropolitan statistical areas (Table 5.6). Again, μSAs correspond to UCs of at least 10,000 residents but less than 50,000. It is important to explore μSAs because they represent smaller, urbanizing areas across the United States. The average BDI score for all μSAs in the U.S. is −21.89 with a standard deviation of 31.34. Again, a negative average value for the BDI is not cause for alarm; even with a negative BDI

Table 5.6 Broadband deployment index: μSAs leaders and laggards, December 2010

Rank	μSAs	BDI	Rank	μSAs	BDI
1	Marion, IN	82.29	377	Port Lavaca, TX	−100
2	New Castle, IN	77.13	376	Price, UT	−100
3	Logansport, IN	76.53	375	Crescent City, CA	−100
4	Greensburg, IN	70.84	374	Sonora, CA	−97.01
5	Decatur, IN	62.67	373	Cornelia, GA	−95.23
6	Plymouth, IN	56	372	Del Rio, TX	−94.59
7	Mitchell, SD	48.48	371	Lake City, FL	−92.95
8	Augusta-Waterville, ME	43.34	370	Liberal, KS	−92
9	Barre, VT	40.59	369	Eureka-Arcata-Fortuna, CA	−91.86
10	Prineville, OR	40.38	368	Middlesborough, KY	−90.85
11	Watertown, SD	38.2	367	Clewiston, FL	−90.85
12	Bennington, VT	35.55	366	Hannibal, MO	−90.32
13	Yankton, SD	32.77	365	Susanville, CA	−88.58
14	Rutland, VT	31.39	364	Taos, NM	−88.15
15	Hood River, OR	31.14	363	Vernal, UT	−88.15
16	The Dalles, OR	31.02	362	Arcadia, FL	−88.15
17	Aberdeen, SD	30.91	361	Gallup, NM	−87.28
18	Grenada, MS	29.96	360	Payson, AZ	−86.92
19	Clarksdale, MS	29.68	359	Williston, ND	−86.2
20	Frankfort, IN	29.12	358	Grants, NM	−85.99

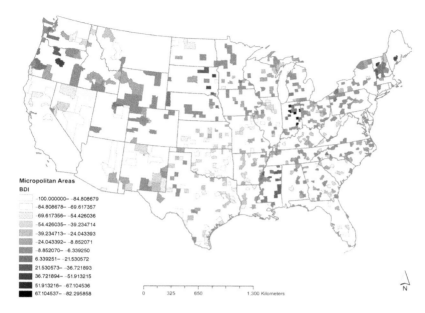

Figure 5.9 Broadband deployment index for micropolitan areas, December 2010.

score, broadband is still available in the vast majority of these communities, but the palette of choices is often reduced when compared to communities that have a strongly positive BDI. From a geographic perspective, the distribution of leaders mimics that of the MSAs and UAs, with many of the μSAs in Indiana, Oregon and Mississippi achieving relatively high BDI scores (Figure 5.9). The μSA laggards correspond to some of the most remote small cities in the United States. This type of geographic isolation, combined with relatively low populations, increases the expense for providers to provide service and decreases their abilities to see a significant return on investment. As a result, most of these communities are subject to monopolies, at least where wireline broadband providers are concerned. Consider, for example, Port Lavaca, TX, which is served by a single cable broadband provider. The community is located midway between Corpus Christi (81 miles [130 km] away), TX and Houston, TX (127 miles [204 km] away) on Lavaca Bay. With just over 21,000 residents in the entire county (Calhoun), the broadband market is not particularly robust. Although peer communities such as Liberal, KS (BDI = −92) fair somewhat better on the BDI, it is still extremely remote (142 miles [229 km] to Amarillo, TX and 211 miles [340 km] to Wichita, KS), and is home to only 20,000 residents. Once again, these conditions often present a poor environment for attracting multiple wireline broadband providers.

Putting it all together

The exploratory spatial data analysis presented in this chapter highlights three major trends in the spatial distribution of wireline broadband provision for the United States. First, it is clear that MSAs exhibit the highest concentration of broadband providers. However, this depth of provision is not necessarily focused on the leading MSAs only (e.g. New York, Chicago and, Los Angeles). While the largest urban corridors such as Megalopolis and the I-25 Front Range corridor in Colorado are dense with wireline options, many smaller MSAs can be considered part of the broadband core for December 2010. This includes locations throughout Indiana, Mississippi and Oregon, all of which display a robust array of wireline providers. This suggests that broadband competition in the core regions is a reality. At the same time, although many smaller urban areas now have a competitive broadband market, this is not the case for many rural and remote communities. Even though broadband is available in these locales, two or fewer providers are the norm in the periphery and many communities are subject to local monopolies. In short, the urban/rural divide still exists, but it is no longer a question of the "have" and "have-nots"; rather, the urban/rural divide has morphed into a *privileged* or *less-privileged* situation. Virtually all communities in the U.S. have broadband. Privileged broadband communities are competitive, home to multiple wireline (and wireless) providers, have a high quality of service and lower prices and are often home to broadband providers that cater strictly to the needs of local firms and businesses (e.g. dedicated fiber or enhanced broadband networks). These privileged communities are also the first locations to receive infrastructure upgrades or benefit from the rollout of new broadband technologies. Less-privileged broadband communities may have fewer choices in providers, spotty wireless broadband coverage (especially on cellular networks), are located further from central locations or major interstate highway corridors, experience lower levels of QoS, higher prices and fewer network upgrades and/or new network technologies. This is reflected, geographically, in Figures 5.3 – 5.6.[9]

A second major trend uncovered in this chapter is the presence of intra-metropolitan heterogeneity in broadband provision. Granted, it should not surprise readers that there are differences within MSAs, however, the geographic dimensions of these differences are interesting. In particular, the spatial gradients in broadband provision appear to be very strong between urban centers and their surrounding suburban and exurban communities. For the purposes of this chapter, the BDI is used as the exploratory tool for detecting the differences between each metropolitan area. However, a non-parametric approach, kernel density analysis, is used to illustrate the spatial decay of broadband provision density for a case study of the Baltimore, MD region. In this case, the results suggest that pockets of stronger broadband provision can exist outside the core city for many metropolitan areas, but the overall decay between the Baltimore UA and more distant regions of the MSA is significant. Even with these provocative results, it is clear that more work is needed to determine how and why these types of gradients exist. One potentially fruitful avenue for investigation is to explore the geographic distribution of firms that specialize in provision of broadband for medium and large-sized businesses. If these providers are clustering in large UAs, their presence may be contributing to the steep provision gradients as one moves toward more suburban locales.

Finally, this chapter uncovered major variations in broadband deployment between micropolitan communities. In addition to the leading μSAs being located closer to larger metropolitan regions than their lagging counterparts, leading μSAs were often located in states that were strongly associated with the broadband core (e.g., Indiana and Oregon.). Lagging μSAs were not only geographically remote, but often bordered on large tracts of undeveloped land, such as national forests, deserts, mountainous regions or coasts. In effect, both their relative isolation and the uninhabited proximal regions suppress the deployment of broadband because of its associated expense and the lack of any economy of scale. Further, because the populations of these communities are under 50,000, the local broadband market is already small and ROI would be limited. That said, it is important to remember that virtually all μSAs have broadband, but the presence of multiple providers in these communities is rare, which is why the BDI is low for many of these locales.

Notes

1 There are variations in subscribership among the states, with the District of Columbia being the lowest at 91% and Colorado the highest at 98.8%. There are also income-related variations. Telephone penetration rates for households with income below $15,000 are 92.8%, while subscription rates exceed 98% for households with incomes over $50,000.

2 Although this is not the most recent release of the NBM data, this information will provide a sufficient snapshot of the broadband landscape in the U.S. Please note that the authors have taken great care to fine-tune these data, mitigate errors and ensure that both small and large blocks are included in the analysis. As detailed in Chapter 4, subsequent versions of the NBM potentially mix 2000 and 2010 block geographies, creating an extremely unstable environment for spatial statistical analysis.

3 Recall that this version of the NBM is based on Census 2000 blocks. To ensure accuracy, this chapter uses Census 2000 block groups for aggregation.

4 For all the readers who are geography geeks, the Census Bureau makes connections between outlying densely settled territories with the main body of the UA or UC for tabulation purposes. "A 'hop' provides a connection from one urban area core to other qualifying urban territory along a road connection of 0.5 miles or less in length; multiple hops may be made along any given road corridor. This criterion recognizes that alternating patterns of residential development and non-residential development are a typical feature of urban landscapes. A 'jump' provides a connection from one urban area core to other qualifying urban territory along a road connection that is greater than 0.5 miles, but less than or equal to 2.5 miles in length; only one jump may be made along any given road connection. The jump concept has been part of the urbanized area delineation process since the 1950 Census, providing a means for recognizing that urbanization may be offset by intervening areas that have not yet developed. The Census Bureau changed

the maximum jump distance from 1.5 miles to 2.5 miles with the Census 2000 criteria" (United States Census Bureau, 2012, FAQ).

5 Readers familiar with autocorrelation statistics might suspect that generating a Voronoi tessellation could mitigate the contiguity problem. In many cases this is true, but for this application it would create geometric borders for Key West that are contiguous to portions of Florida for which no true spatial relationship exists (e.g., block groups near Naples, FL). Readers might also suggest a distance-based contiguity matrix, but this also has problems because of the large deviations in block group size (and associated centroid point distances) between urban and rural areas.

6 This index can also be structured to capture differences in urbanized areas such as UAs or UCs.

7 For a more detailed discussion on the relative advantages and disadvantages associated with the use of different metrics for measuring broadband presence, see Mack (2014b).

8 It is interesting to note that the broadband landscape for Fishers, IN has changed very little between 2010 and 2013. There are still six unique providers (adjusting the 2010 count for the MegaPath and Covad merger). This is likely the case for most broadband core regions. However, as noted previously in this chapter and Chapters 4 and 5, the most recent releases of the NBM are not particularly stable and we are hesitant to use them for analysis.

9 Although issues associated with pricing, QoS and many of the other facets detailed in the "privileged and less-privileged" hierarchy were not empirically addressed in this chapter, recent work in the field, both in the U.S. and elsewhere, supports these characterizations (see Grubesic, 2006, 2008, 2014; Prieger, 2013; Townsend et al., 2013; Riddlesden & Singleton, 2014).

References

American Recovery and Reinvestment Act [ARRA]. (2009). Retrieved from http://en.wikisource.org/wiki/American_Recovery_and_Reinvestment_Act_of_2009

Andronico, G., Ardizzone, V., Barbera, R., Becker, B., Bruno, R., Calanducci, A., . . . & Scardaci, D. (2011). e-Infrastructures for e-Science: a global view. *Journal of Grid Computing*, *9*(2), 155–184.

Anselin, L. (1995). Local indicators of spatial association—LISA. *Geographical Analysis*, *27*(2), 93–115.

Anselin, L., Varga, A., & Acs, Z. (1997). Local geographic spillovers between university research and high technology innovations. *Journal of Urban Economics*, *42*, 422–448.

Belleflamme, P., Picard, P., & Thisse, J.F. (2000). An economic theory of regional clusters. *Journal of Urban Economics*, *48*(1), 158–184.

Benton Foundation. (2009). *96% of US households have telephones*. Retrieved from https://www.benton.org/node/41269

Bergman, E.M., & Feser, E.J. (1999). Industrial and regional clusters: Concepts and comparative applications. In *Web book in regional science*. West Virginia University, USA: Regional Research Institute.

Cairncross, F. (2001). *The death of distance: How the communications revolution is changing our lives*. Harvard Business Press.

Census (2010). *Jackson County Quick Facts*. Retrieved from http://quickfacts.census.gov/qfd/states/17/17077.html

Charles, D., Bradley, D., Chatterton, P., Coombes, M., & Gillespie, A. (1999). *Core cities: Key centres for regeneration*; Synthesis Report. University of Newcastle upon Tyne, Centre for Urban and Regional Development Studies.

Compaine, B.M. (Ed.). (2001). *The digital divide: Facing a crisis or creating a myth?* MIT Press.

Federal Communications Commission [FCC]. (2010a). *Connecting America: The National Broadband Plan*. Retrieved from http://download.broadband.gov/plan/national-broad band-plan.pdf

Federal Communications Commission [FCC]. (2010b). *Sixth broadband deployment report*. Retrieved from http://transition.fcc.gov/Daily_Releases/Daily_Business/2010/db0720/FCC-10-129A1.pdf

Federal Communications Commission [FCC]. (2012). *Eighth broadband deployment report*. Retrieved from https://apps.fcc.gov/edocs_public/attachmatch/FCC-12-90A1.pdf

Federal Register. (2011). *Criteria used to define urban areas for the 2010 Census*. Retrieved from https://www.census.gov/geo/reference/frn.html

Flamm, K., & Chaudhuri, A. (2007). An analysis of the determinants of broadband access. *Telecommunications Policy*, *31*, 312–326.

Forman, C., Goldfarb, A., & Greenstein, S. (2005). How do industry features influence the role of location on Internet adoption? *Journal of the Association for Information Systems*, *6*(12), 383–408.

Friedmann, J., & Miller, J. (1965). The urban field. *Journal of the American Institute of Planners*, *31*(4), 312–320.

Goldfarb, A., & Prince, J. (2008). Internet adoption and usage patterns are different: Implications for the digital divide. *Information Economics and Policy*, *20*(1), 2–15.

Gordon, I.R., & McCann, P. (2000). Industrial clusters: complexes, agglomeration and/or social networks? *Urban studies*, *37*(3), 513–532.

Graham, S. (1999). Global grids of glass. *Urban Studies*, *36*(5/6), 929–949.

Graham, S., & Marvin, S. (1996). *Telecommunications and the city*. London: Routledge.

Greene, F.J., Tracey, P., & Cowling, M. (2007). Recasting the city into city-regions: Place promotion, competitiveness benchmarking and the quest for urban supremacy. *Growth and Change*, *38*(1), 1–22.

Greenstein, S. (2007). *Data constraints & the Internet economy: Impressions & imprecision*. NSF/OECD meeting on Factors Shaping the Future of the Internet. Retrieved from http://www.oecd.org/dataoecd/5/54/38151520.pdf

Greenstein, S., & Prince, J. (2006). *The diffusion of the Internet and the geography of the digital divide in the United States (No. w12182)*. National Bureau of Economic Research.

Grubesic, T.H. (2003). Inequities in the broadband revolution. *The Annals of Regional Science*, *37*(2), 263–289.

Grubesic, T.H. (2006). A spatial taxonomy of broadband regions in the United States. *Information Economics and Policy*, *18*(4), 423–448.

Grubesic, T. H. (2008). The spatial distribution of broadband providers in the United States: 1999–2004. *Telecommunications Policy*, *32*(3), 212–233.

Grubesic, T.H. (2010). Efficiency in broadband service provision: A spatial analysis. *Telecommunications Policy*, *34*(3), 117–131.

Grubesic, T.H. (2012a). The US National Broadband Map: Data limitations and implications. *Telecommunications Policy*, *36*(2), 113–126.

Grubesic, T.H. (2012b). The wireless abyss: Deconstructing the US National Broadband Map. *Government Information Quarterly*, *29*(4), 532–542.

Grubesic, T.H. (2015). The broadband provision tensor. *Growth and Change*. DOI: 10.1111/grow.12083

Grubesic, T.H., & Horner, M.W. (2006). Deconstructing the divide: Extending broadband xDSL services to the periphery. *Environment and Planning B: Planning and Design*, *33*(5), 685.

Grubesic, T.H., & Murray, A.T. (2002). Constructing the divide: Spatial disparities in broadband access. *Papers in Regional Science*, *81*(2), 197–221.

Grubesic, T.H., & Murray, A.T. (2004). Waiting for broadband: Local competition and the spatial distribution of advanced telecommunication services in the United States. *Growth and Change*, *35*(2), 139–165.

Grubesic, T.H., Matisziw, T.C., & Murray, A. (2010). Market coverage and service quality in digital subscriber lines infrastructure planning. *International Regional Science Review*, *34*(3), 368–390.

Haffner, J.H., Gonzalez, D.G., Terracina-Hartman, C., Bienkowski, B., Myers, M., Kanthawala, S., . . . & Carloni, G. (2013). Public, private and non-profit interventions in California's digital divide: A case for thoughtful action. *International Journal of Technology, Knowledge & Society*, *9*(4).

Hales, B., Gieseke, J., & Vargas-Chanes, D. (2000). Telecommunications and economic development: Chasing smokestacks with the internet. In Korsching, P.F., Hipple, P.C., & Abbott, E.A., (Eds.), *Having all the right connections: Telecommunications and rural viability* (pp. 257–276). Westport, CT: Praeger.

Hoffman, D.L., & Novak, T.P. (1998). Bridging the digital divide: The impact of race on computer access and internet use. *Science*, *280* (April 17), 390–391.

John, C.H., & Pouder, R.W. (2006). Technology clusters versus industry clusters: Resources, networks, and regional advantages. *Growth and Change*, *37*(2), 141–171.

Johnson, D.K., Siripong, A., & Brown, A.S. (2006). The demise of distance? The declining role of physical proximity for knowledge transmission. *Growth and Change*, *37*(1), 19–33.

Kelsall, J. E., & Diggle, P.J. (1995). Non-parametric estimation of spatial variation in relative risk. *Statistics in Medicine*, *14*(21–22), 2335–2342.

Kenney, M. (Ed.). (2000). *Understanding Silicon Valley: The anatomy of an entrepreneurial region*. Stanford University Press.

Kloch, C., Petersen, E.B., & Madsen, O.B. (2011). Cloud based infrastructure, the new business possibilities and barriers. *Wireless Personal Communications*, *58*(1), 17–30.

Kokomo Perspective (2013). Kokomo is the eighth fastest growing city in the U.S. Retrieved from http://tinyurl.com/m2nq39l

Langford, M., & Unwin, D.J. (1994). Generating and mapping population density surfaces within a geographical information system. *The Cartographic Journal*, *31*(1), 21–26.

LaRose, R., Strover, S., Gregg, J.L., & Straubhaar, J. (2010). The impact of rural broadband development: Lessons from a natural field experiment. *Government Information Quarterly*, *28*(1), 91–100.

Livingstone, S., & Helsper, E. (2007). Gradations in digital inclusion: Children, young people and the digital divide. *New Media & Society*, *9*(4), 671–696.

Mack, E.A. (2014a). Broadband and knowledge intensive firm clusters: Essential link or auxiliary connection? *Papers in Regional Science*, *93*(1), 3–29.

Mack, E.A. (2014b). Businesses and the need for speed: The impact of broadband speed on business presence. *Telematics and Informatics*, *31*(4), 617–627.

Mack, E.A., & Grubesic, T. H. (2014). US broadband policy and the spatio-temporal evolution of broadband markets. *Regional Science Policy & Practice, 6*(3), 291–308.

Mack, E.A., & Rey, S.J. (2014). An econometric approach for evaluating the linkages between broadband and knowledge intensive firms. *Telecommunications Policy, 38*(1), 105–118.

Mack, E.A., Anselin, L., & Grubesic, T.H. (2011). The importance of broadband provision to knowledge intensive firm location. *Regional Science Policy and Practice, 3*(1), 17–35.

MegaPath (2010). *MegaPath, Covad, Speakeasy merger closes*. Retrieved from http://www.megapath.com/about/press-releases/megapath-covad-speakeasy-merger-closes/

Office of Management and Budget [OMB]. (2010). *Standards for delineating metropolitan and micropolitan statistical areas*. Retrieved from http://tinyurl.com/mud5o97

Organization for Economic and Cooperative Development [OECD]. (2013). *OECD broadband portal*. Retrieved from http://www.oecd.org/sti/broadband/oecdbroadband portal.htm

Oyana, T.J. (2011). Exploring geographic disparities in broadband access and use in rural southern Illinois: Who's being left behind? *Government Information Quarterly, 28*(2), 252–261.

Parkison, D. (2011). *Kokomo forecast 2012*. Retrieved from http://www.ibrc.indiana.edu/ibr/2011/outlook/kokomo.html

Petruzzelli, A.M. (2011). The impact of technological relatedness, prior ties, and geographical distance on university–industry collaborations: A joint-patent analysis. *Technovation, 31*(7), 309–319.

Pew Research Center. (2013). *Home broadband*. Retrieved from http://www.pewinternet.org/2013/08/26/home-broadband-2013/

Prieger, J., & Lee, S. (2008). Regulation and the deployment of broadband. In Dwivedi, Y.K. (Ed.), *Handbook of research on global diffusion of broadband data transmission* (pp. 278–303). Igi Global.

Riddlesden, D., & Singleton, A.D. (2014). Broadband speed equity: A new digital divide? *Applied Geography, 52*, 25–33.

Rosston, G.L., Savage, S.J., & Waldman, D.M. (2010). Household demand for broadband Internet in 2010. *The BE Journal of Economic Analysis & Policy, 10*(1), 79.

Saxenian, A. (1991). The origins and dynamics of production networks in Silicon Valley. *Research Policy, 20*(5), 423–437.

Saxenian, A. (1996). Inside-out: Regional networks and industrial adaptation in Silicon Valley and Route 128. *Cityscape: A Journal of Policy Development and Research, 2*(2), 41–60.

Silicon Valley Index [SVI]. (2014). *Initial public offerings*. Retrieved from http://www.siliconvalleyindex.org/svi/index.php/economy/entrepreneurship

Smith, S. (2014). Kokomo economic growth has continued. *Kokomo Tribune*. Retrieved from http://tinyurl.com/l3wmsxx

Steinfield, C., Scupola, A., & López-Nicolás, C. (2010). Social capital, ICT use and company performance: Findings from the Medicon Valley Biotech Cluster. *Technological Forecasting and Social Change, 77*(7), 1156–1166.

Sylvester, D.E., & McGlynn, A.J. (2010). The digital divide, political participation, and place. *Social Science Computer Review, 28*(1), 64–74.

Townsend, L., Sathiaseelan, A., Fairhurst, G., & Wallace, C. (2013). Enhanced broadband access as a solution to the social and economic problems of the rural digital divide. *Local Economy, 28*(6), 580–595.

United States Census Bureau. (2012). *What are hops and jumps in the urban area delineation criteria?* Retrieved from https://ask.census.gov/faq.php?id=5000&faqId=5963

Van Dijk, J., & Hacker, K. (2003). The digital divide as a complex and dynamic phenomenon. *The Information Society, 19*(4), 315–326.

Wheeler, J.O., Aoyama, Y., & Warf, B. (Eds.). (2000). *Cities in the telecommunications age: The fracturing of geographies*. Psychology Press.

Whitacre, B.E. (2010). The diffusion of Internet technologies to rural communities: A portrait of broadband supply and demand. *American Behavioral Scientist, 53*(9), 1283–1303.

Whitacre, B.E., & Mills, B.F. (2007). Infrastructure and the rural-urban divide in high-speed residential Internet access. *International Regional Science Review, 30*(3), 249–273.

6 The Broadband-Business Nexus

As detailed in the previous five chapters, broadband often, although not exclusively, serves as a gateway technology, fundamentally transforming the way both people and businesses operate (Czernich, Falck, Kretschmer, & Woessmann, 2011). In particular, the Internet has dramatically enhanced operational efficiency, improved the visibility of businesses, opened up new business opportunities and encouraged innovation through collaboration (Wright, 1992; Horvath, 2001; Kloch, Petersen, & Madsen, 2011; Andronico et al., 2011). As a result, the Internet has quickly become a fundamental component of global critical infrastructure. Most, if not all sectors of the global economy including energy, banking and finance, and homeland security are completely reliant on broadband internet connections for timely and efficient communications (Murray & Grubesic, 2007). Because of the unparalleled reliance of people, businesses and governments on internet-enabled technologies, vast amounts of time, energy and money have been directed toward rolling out internet infrastructure around the world.

The increasing reliance of businesses and individuals on information technologies, such as broadband, has prompted widespread scientific inquiry into the nature and sources of gaps in the provision, adoption and use of information and communications technologies (ICTs). As detailed in Chapter 5, the bulk of this research focuses on the digital divide from a residential market perspective. Results suggest that many disparities in access remain in the United States and other countries around the globe, some of which can be connected to local demographic structure, socio-economic status and educational attainment (Hoffman & Novak, 1999; Gabel & Kwan, 2001; Norris, 2001; Grubesic, 2004; Grubesic & Murray, 2004; Chakraborty & Bosman, 2005; Mossberger, Tolbert, & McNeal, 2006; LaRose, Gregg, Strover, Straubhaar, & Carpenter, 2007; Prieger & Hu, 2008; García, Thapa, & Niehaves, 2014; Nishida, Pick, & Sarkar, 2014; van Deursen & Van Dijk, 2014).

Surprisingly, the bias of internet-related research toward residential adoption and use means that the impacts of broadband on businesses remains relatively understudied. In part, this can be attributed to the complexities associated with broadband and businesses. Not only is the relationship between businesses and telecommunications technologies relevant (e.g., access, platform availability and use patterns), but the interactions *between* businesses facilitated by ICTs require a deeper understanding. Further, there is a case to be made for uncovering the dynamic relationships between businesses and broadband telecommunication providers, particularly as it relates to supply and demand issues for bandwidth, pricing and bundled services. Thus, although some research has begun to explore the linkages between business presence and broadband infrastructure (Kandilov & Renkow, 2010; Mack, Anselin, & Grubesic, 2011), more work is required to understand the intricacies underlying this relationship, especially given the expanding impact of telecommunications infrastructure on the regional business environment (Nucciarelli, Castaldo, Conte, & Sadowski, 2013; Ramsay, 2013; Whitacre, Gallardo, & Strover, 2014).

The purpose of this chapter is to develop and describe a conceptual framework, the Broadband-Business Nexus (BBN), which clarifies the many potential linkages that exist between broadband and businesses – all of which take place in a dynamic and constantly evolving landscape of local and regional broadband markets. Although this framework is designed to function as a conceptual tool for researchers, policymakers and economic development officials, disentangling the messy web of interactions in these markets cannot be fully accomplished using publicly accessible data. Many of the potential relationships detailed in this chapter are extremely difficult to identify from the outside, looking in. However, this is a challenge common to all ICT research, particularly when private companies (e.g., Verizon and Comcast) are unwilling to share their data. That said, the framework developed in this chapter provides enough flexibility to allow for such connections to be acknowledged even if their exact manifestation is not completely understood.

In the next section, and before outlining the details of the BBN, this chapter reviews the relevant literature concerning information and communications technologies and firms. This overview covers broadband availability, adoption, use, productivity, firm location and the impact of these linkages on the economic development prospects of regional economies. After this overview of the relevant literature, we introduce the BBN.

Availability of infrastructure

The availability of ICT infrastructure, such as broadband, is an important consideration for firms and their location decisions (Nucciarelli et al., 2013; Ramsay, 2013; Whitacre et al., 2014). However, the core dimensions of ICT availability are nuanced, especially from a geographic perspective. Also, depending on firm type, size and scope of operations, their associated ICT needs can be highly varied. Many of these details associated with ICT availability are discussed in previous chapters, but there are two facets worth reiterating here. First, it is important to remember that although the historical context of the Internet and its development remains firmly rooted in the public domain (Abbate, 1999), the U.S. backbone infrastructure was handed over to the private sector in 1995. Since this date, private telecommunications firms have had the responsibility of deploying this infrastructure to people and businesses (Downes & Greenstein,

2007). Recent market analysis suggests that private firms have invested $1.6 trillion in capital on broadband infrastructure since 1996, with $68 billion invested in 2012 alone (Brogan, 2013). Simultaneously, consumer expenditures on broadband are massive. For example, consumers spent $54 billion on wireless services and $49.6 billion on terrestrial broadband services during 2013 (Brogan, 2013). Of note is that these infrastructure investments and associated expenditures are not geographically ubiquitous. Investments have proved to be spatially inequitable, with providers cherry-picking the most profitable markets for infrastructure rollouts, leaving many rural and remote areas with fewer provider choices, higher prices and lower quality of service (QoS) when compared to urban and suburban locales (Grubesic & Murray, 2002; Grubesic, 2006; Prieger, 2013; Townsend, Sathiaseelan, Fairhurst, & Wallace, 2013). There are exceptions, of course, where highly localized pockets of limited broadband deployment manifest in large metropolitan areas (Grubesic, 2006). These areas are often home to impoverished minority populations that broadband providers deem unprofitable to service.

Although disparities in availability are not uncommon for new technologies, the manner in which these inequities have evolved for broadband proves particularly challenging for conducting empirical analysis, crafting meaningful policy and encouraging economic development. Broadband availability is no longer a matter of the haves and have-nots (Grubesic, 2015). Again, gaps in broadband availability have become more nuanced as the technology matures. Today, the broadband divide is a complex and multi-faceted issue, challenging existing business models, business operations and forcing policymakers to craft more innovative solutions to meet community needs. Disparities in access now include a host of issues that include platform choice (e.g., cable, DSL and wireless), tiered speed/capacity plans and service quality, amongst other things (Riddlesden & Singleton, 2014; Grubesic, 2015). These disparities are particularly salient for firms that use ICTs intensively in their business processes. Prior work on ICTs and firm presence suggests that heterogeneities in ICT availability and speed can affect a firm's location decision (Sohn, Kim, & Hewing, 2003; Mack, 2014). Further, the linkages between broadband and firms can be highly variable across metropolitan areas (Mack et al., 2011; Mack & Faggian, 2013), as can the impact of broadband loans on business presence (Kandilov & Renkow, 2010).

Information and communications technologies (ICTs) and firm location

A large body of the work on ICT adoption and its associated impacts, particularly the research related to broadband, focuses on individuals and/or households rather than businesses. However, this trend in the literature is changing as more researchers take on the task of evaluating the linkages between broadband and businesses (Mack & Grubesic, 2009; Kandilov & Renkow, 2010; Kolko, 2010; Mack et al., 2011; Mack, 2014; Mack & Rey, 2014) and broadband and entrepreneurial activity (Heger, Veith, & Rinawi, 2011; Parajuli & Haynes, 2012). Nevertheless, much more work is needed to better understand *if* and *how* broadband changes the landscape of business location.

Early theoretical work offered several hypotheses about how and why innovations in information and communication technologies *may* affect business location decisions thereby transforming the competitive position of regional economies (Goddard & Gillespie, 1986; Gillespie & Williams, 1988; Capello & Nijkamp, 1996; Röller & Waverman, 2001). This is not to say that the availability of infrastructure makes places more competitive, rather the presence or absence of critical infrastructure, such as broadband for businesses, may be a *differentiating* location factor. Recent work suggests that telecommunications infrastructure has a profound impact on regional business climates (Nucciarelli et al., 2013; Ramsay, 2013; Whitacre et al., 2014). Locational preferences are no longer a simple trade-off between production and transportation costs at the intra-national level, but must now include access to global markets, the transactions costs of information transmission, and the frequency of face-to-face interactions with local and global contacts (McCann & Sheppard, 2003).

The consensus of much early theoretical work was that ICTs would impact business location in one of three ways: 1) complete dispersal, 2) reinforced concentration or 3) mixed patterns of concentration and dispersal (Audirac, 2005; Mack & Grubesic, 2009). These paradigms are rooted in a larger debate about the fate of cities in an informational economy (Salomon, 1996). The first prediction about firm location is related to the death of distance concept: the availability of high-quality telecommunications infrastructure will result in a mass departure of firms from expensive and congested central city locations (Negroponte, 1995; Cairncross, 1997), with firms relocating to less expensive suburban locales to avoid the high rents and traffic congestion associated with downtown areas (Kutay, 1988a, 1988b). Of course, this process of dispersion is rooted in three key assumptions. The first is that communication via ICTs is perfectly substitutable for the agglomerative benefits of central city locations. For example, in this paradigm, technologies such as video conferencing function as a substitute for face-to-face interaction (Negroponte, 1995; Moss, 1998; Steinfield, 2004). The second important assumption of this paradigm is that ICT infrastructure is ubiquitously available throughout regions. Subsequently, firms will be free to choose the location that minimizes the costs of their business activities (Salomon, 1996). A third assumption is that telecommunications can also substitute for transportation (Kraemer, 1982; Mokhtarian, 1990; Salomon, 1996). This assumption effectively frees workers from a daily commute, allowing them to work from home or an alternative location.

Although arguments for dispersal are compelling, there are some fundamental problems with its core assertions. The first problem is the inability of telecommunications to substitute for face-to-face interactions. Cities both reduce travel costs and greatly facilitate face-to-face interactions (Gaspar & Glaeser, 1998). This is particularly true for non-standard and/or complex exchanges that require face-to-face collaboration or communication (Leamer & Storper, 2001; McCann & Shefer, 2004). Thus, the virtual connections made possible by

ICTs are likely a complement to rather than a substitute for in-person transactions (Gaspar & Glaeser, 1998; Moss, 1998). This is not unlike e-commerce, which is a strong complement to physical retail spaces (Steinfield, 2004), but not a complete substitute. Cities also continue to be important for facilitating in-person, non-standard (e.g., highly technical and sometimes social) transactions despite the advent of today's advanced, knowledge economy (Pons-Novell & Viladecans-Marsal, 2006).

A second paradigm revolves around the concentration of business activities due to the uneven distribution of ICT resources. This school of thought also operates on a handful of basic assumptions. First, much of the early work concerning the spatial organization of ICT infrastructure suggested that it was unevenly distributed, with the bulk of its critical components located in a handful of large cities (Grubesic & O'Kelly, 2002; O'Kelly & Grubesic, 2002; Duffy-Deno, 2003). This meant that the highest bandwidth connections and the most reliable services were only available in a limited number of locales, effectively disrupting the notion of infrastructure ubiquity and the freedom for firms to locate anywhere they pleased. A second key assumption of the concentration paradigm is that that the centrifugal forces of agglomeration economics will continue to draw firms to central locations (Atkinson, 1998). The opportunities for knowledge spillovers and the thick labor markets present in central cities are benefits not necessarily offered by peripheral, less expensive locales. Thus, the impact that telecommunications will have on firm location is ultimately a question of the importance of agglomerative benefits to the firm.

Many of the arguments for concentration are persuasive; cities are, after all, a spatial strategy for minimizing transport costs, travel times and infrastructure expenses by co-locating services and facilities in a small area that is highly accessible to both residents and/or workers. However, infrastructure, businesses and the urban populations in the United States have been gravitating to more peripheral suburban and exurban locations since the 1950s (Jackson, 1985; Mieszkowski & Mills, 1993). This build-out has resulted in a relatively homogenous landscape for telecommunications infrastructure. Although teledensity is lower in suburban areas (Grubesic, 2003) and the limitations of distance sensitive technologies, such as xDSL, are exposed in these less-central locations (Grubesic & Murray, 2002; Grubesic & Horner, 2006; Grubesic, Matisziw, & Murray, 2011), high-quality fiber and cable-based technologies are readily available for both residential and commercial deployment (Mitcsenkov Paksy, & Cinkler, 2011). As a result, although telecommunications infrastructure is still far from ubiquitous, for most suburban locales, it is virtually identical to what is available in the nearby central city even if the "thickness" of the infrastructure dissipates somewhat in the suburban locales.[1]

Finally, many scholars suggest ICTs will have a heterogeneous impact on firm location (Audirac, 2005; Arbore & Ordanini, 2006; Mack et al., 2011). The foundational argument of this school of thought is that ICTs provide firms with the opportunity to relocate to peripheral locations should they decide it is beneficial to their business processes (Kutay, 1986). Thus, telecommunications are a necessary but not sufficient condition for regional development and/or relocation (Capello, 1994). Adherents to this school of thought suggest several reasons for this hetero-geneous impact. For example, one characteristic of firms is the skill level of their business processes. Telecommunications have enabled firms to decouple basic back office activities to lower cost, less-central locations (Warf, 1989; Youtie, 2000; Graham & Marvin, 2001). Another consideration is industry-specific char-acteristics (Atkinson, 1998; Moss, 1998; Audirac, 2005). Specifically, the degree that firms require proximity to suppliers, other firms and their customers will vary by industry (Atkinson, 1998).

This is clearly the most balanced of the three schools of thought. Not only does it suggest that ICTs are mediated by the specific contexts of individual indus-tries, the contextual elements of the local telecommunications landscape are also factors. For example, if the appropriate ICT platforms, capacity and/or QoS are not uniformly distributed in a region, firms will be more likely to gravitate to sub-regions where the ICTs can accommodate their needs. Thus, some firms will select central city locations while others will select more suburban locations.

Firm adoption and use of information technology

Broadband availability and the impacts of ICTs on firm location decisions are only two of the aspects of the business/broadband connection that should be con-sidered. Another important piece of the puzzle involves a consideration of the adoption and intensity of broadband use by businesses. While business adoption and use of broadband remain comparatively understudied compared to residential adoption and use, the broader literature discussing information and communi-cations technologies and firms offers some clues about this relationship (Mack et al., 2011).

Firm adoption of ICTs is fueled by a multitude of factors, including geogra-phy, industry membership and business size. For example, at the confluence of all three facets, the early work on dial-up internet adoption (Forman, Goldfarb, & Greenstein, 2003, 2005a) suggests that the geography of establishments has an impact on adoption rates. Specifically, when controlling for industry type and size these studies found that dial-up adoption was lower in small metropolitan and rural areas. Other studies have also found that the geographic location of both broadband *and* businesses plays a role in this relationship. For example, Mack and Grubesic (2009) uncovered an urban bias to both broadband and knowledge-intensive firms in Ohio, which may explain the lack of firm decentralization from large metropolitan areas throughout the state.

The findings of Forman et al. (2003, 2005b) and Mack and Grubesic (2009) also suggest that industry membership can play an important role in explaining *how* and *why* businesses adopt and use ICTs. For example, financial services firms (NAICS 52) are widely noted to be some of the heaviest users of telecommunica-tion services (Korek & Olszewski, 1981; Warf, 1989). Information (NAICS 51), Professional, Scientific, and Technical Services (NAICS 54), and Management of Companies and Enterprises (NAICS 55) are also well-known to be knowledge- and technology-intensive industries (Mudambi, 2008), which often increases their ICT adoption rates compared to other sectors (Shiels, McIvor, & O'Reilly, 2003). This intensity of use is not only important for the industries immediately involved but also the telecommunications companies that provide service to these sectors. For example, financial services firms are heavy investors in new telecommunications technologies (Warf, 1989). In this sense, the local presence of businesses in particular industries, such as finance, generates significant demand for telecommunications services and spurs infrastructure investments even when residential broadband demand is insufficient to stimulate capital expenditures.

However, as mentioned earlier, there are nuances to this landscape. For instance, where the global services industry is concerned, particularly finance, there is a need to consider the uses of telecommunications for these firms. In this regard, the division of telecommunications uses can be subdivided into two categories, *participation* and *enhancement* (Forman et al., 2005a). Participation uses are considered the most basic applications of ICT use and include email, web-browsing and electronic sharing of documents (ibid.). Enhancement uses of the Internet are those that fundamentally change business operations or allow companies to implement and offer new products and services (ibid.).

In this context, enhancement-related uses are expected to increase the productivity of businesses, which has certainly been the case for the global services industries (Warf, 1995). This is not necessarily the case for businesses in other industries, however. The productivity impacts, and thus the reduction in costs that are associated with telecommunications use, are heterogeneous and depend on several factors. These factors include the type of work and the skill level of workers using the technology (Jorgenson, 2001) and the time it takes for the productivity impacts to be realized. Studies about broadband and productivity have found evidence of skills-biased technological change (Mack & Faggian, 2013). This finding is in line with the work of Kolko (2010) who found that a switch from dial-up to broadband only impacted the use of particular internet applications (e.g., music downloads and online shopping) that are not necessarily associated with productive uses of the Internet. A study by Martínez, Rodriguez and Torres (2010), for example, suggests that there is a temporal lag in productivity impacts because it takes time for companies to adapt to the new technology itself, as well as make organizational changes resulting from the technological shift. Combined, this body of research suggests that *who* uses new technologies is just as important as the initial adoption of these technologies.

In terms of business adoption and use, firm size can play an important role. For example, larger firms are more likely to have in-house IT support when compared to smaller firms (Karshenas & Stoneman, 1993; Gibbs & Tanner, 1997; Forman, 2005). This not only impacts the ability of firms to implement new technologies because of differences in IT expertise and resources (Gibbs, 2001), but it also impacts firm awareness about the advantages of using ICTs. In fact, studies have suggested that this lack of awareness makes small firms less likely to adopt ICTs than larger, more informed firms (Center for an Urban Future, 2004). Paradoxically, these differences in ability and awareness are critical to consider for extremely small enterprises that may have just one or two people who own, operate and manage the ICT assets, and who will also perhaps experience the greatest impact in terms of firm visibility from the implementation of IT solutions.

Telecommunications market dynamics

A final consideration worth detailing before introducing the BBN is the nature of telecommunication market dynamics. These dynamics are important because policy, regulatory, technological and industry trends in telecommunications can have a dramatic impact on corporate and residential customers, as well as influencing QoS, pricing and access equity at a local level. Moreover, although this chapter will not delve into issues concerning labor, political and cultural impacts of telecommunications markets, they too add to the complex undercurrents of the telecommunications landscape (Warf, 2003).

At the federal level, the *Telecommunications Act of 1996* (*96 Act*) is approaching its 20th anniversary in 2016 and it remains one of the most important policy development in the modern era. The *96 Act* was the first major overhaul in federal telecommunications policy in sixty years and was structured to encourage a free and competitive market where commercial telecommunication providers would compete for both residential and commercial accounts (Grubesic & Murray, 2004). Embedded within the *96 Act*, Section 706 directed the Federal Communications Commission and all 50 states to encourage the deployment of advanced telecommunications capabilities to all U.S. residents in a reasonable and timely manner. Although the temporal component of this directive was poorly defined, advanced telecommunications capability referred to broadband, which at the time was considered to be 200 kbps transmission speeds for upstream and downstream connections – enough to allow users to "originate and receive high-quality voice, data, graphics and video telecommunications using any technology" (*96 Act*). This was the spark that lit the broadband revolution in the United States. There is a huge literature that details the relative successes and failures of the *96 Act*, and we direct readers to the work of Aufderheide (1999, 2006), Crandall (2005), Frieden (2005) and Cambini and Jiang (2009), as well as Chapter 2 of this book for additional perspective on this legislation and its impact on local, regional and national markets for broadband services.

At the state level, public utility commissions (PUCs) are a key source of telecommunication regulation and policy reform. For example, Kim and Gerber (2005) note that after the breakup of AT&T, the Regional Bell Operating Companies (RBOCs) fell under state jurisdiction and all decisions regarding intra-local access and transport areas (LATA),[2] as well as inter-LATA competition were under state government control. However, this did not release RBOCs from federal regulation. The *96 Act* required RBOCs (and other local incumbents) to lease their facilities to competitive local exchange carriers (CLECs) to promote competition. During the late 1990s and early 2000s, this unbundling of the local copper infrastructure meant that xDSL services could be marketed and sold by CLECs, with almost no infrastructure investment from these competitive carriers. This had profound impacts on local broadband markets (Grubesic & Murray, 2002, 2004; Grubesic, 2004), but effectively came to a close in 2004 with a ruling from the District of Columbia Circuit Court (DCCC, 2004) that limited infrastructure sharing between RBOCs, ILECs and CLECs.

There are other examples of state broadband policy influencing market dynamics. For instance, there are 21 states that guarantee rights-of-way to telecommunication firms (NTIA, 2003).[3] This typically includes the right to trench, place wireless antenna structures or plant poles where necessary to expand and/or maintain the broadband network. However, regardless of guarantees, obtaining rights-of-way for most broadband providers is an expensive, time consuming and often contentious process. As detailed by the NTIA (2003, 1):

> A number of providers, for example, noted that deployment was often slowed by overly burdensome requests for information, lengthy processes for obtaining permits, unreasonable charges for use of the rights-of-way, and undue remediation and maintenance requests. Additionally, several providers noted that the complex patchwork of procedures among localities made installation of facilities across municipal boundaries costly and time-consuming.

Interestingly, there are six states that exempt cable companies from these guaranteed rights-of-way benefits (Wallsten, 2005). One would think that such limitations might inhibit the rollout of broadband infrastructure, and in some cases, perhaps it did. However, Crandall, Hahn, Litan, and Wallsten (2004) suggest that cable broadband grew faster than xDSL because telecommunications companies (e.g., RBOCs) were regulated at the federal level, but cable companies were not. Clearly, this interplay between federal and state regulations is yet another example of the complexity of telecommunication policy in the United States and its impact on local market dynamics.

There are many other examples of regulations or directed programs at the state and local levels that influence broadband provision and market dynamics. For example, USDA Rural Development grants provided $1.4 billion to rural areas for telecommunications projects in 2004. These monies were used to construct, improve or expand telecommunications infrastructure (Wallsten, 2005; Kandilov & Renkow, 2010) in a variety of rural and remote communities. More recently, private projects, such as Google Fiber (Google, 2014), require strong coordination between local city officials, government agencies and regulators to ensure installations are carried out smoothly and adhere to local standards. One must also consider the impacts of telecommunications mergers and acquisitions (Warf, 2003). As was detailed previously, the impending $45 billion dollar merger between Comcast and Time Warner will fundamentally change the landscape of broadband provision in the U.S. and will impact nearly 70 million households and 18 million businesses. This mega-merger is sure to impact pricing, QoS and local competition, although it is difficult to predict the exact outcomes.

Lastly, it is important to reiterate that necessary information about broadband provision, access and service quality is not always available. These information asymmetries exist because the relevant data are considered proprietary and service providers refuse to disclose information about subscribers, QoS metrics and/or pricing due to fears of losing their competitive edge in the market (Greenstein, 2007). Historically, even when information was made available, it was usually done in highly aggregated fashion, such as ZIP codes or Census tract data, although the National Broadband Map is now making higher resolution data available for analysis (Grubesic, 2012). Still, most policymakers, businesses and local economic development officials have incomplete knowledge regarding the types of broadband resources that are available in their region.

In addition to the lack of information on broadband in disaggregate form, the coverage of information that is available is often imperfect as well. For example, while information on availability, pricing and QoS may be available in some locales, it is rarely available for all locations in a region. This imperfect regional information can negatively impact investment decisions on telecommunication infrastructure. Notably, these types of information asymmetries are a common feature in unsuccessful municipal investments in fiber and wireless broadband systems (Gillett, Lehr, & Osorio, 2004; Lehr et al., 2004; Gibbons & Ruth, 2006; Sirbu, Lehr, & Gillett, 2006).

Broadband-Business Nexus

Given the wide range of factors impacting the linkage between broadband technologies, local markets and firms, the development of a conceptual framework that enhances our understanding of these dynamics is critical for crafting telecommunication policies that help facilitate regional development. In fact, the BBN seeks to formalize suspected, yet radically understudied connections between businesses and broadband in the U.S. In short, the BBN facilitates a basic typology of interactions and provides a framework for exploring the interplay between businesses and broadband telecommunication services. Figure 6.1 displays the core components of the BBN (A – D), all of which are required to dissect the complex interactions between businesses and broadband telecommunication services. Specifically, not only does this framework require a consideration of the economic, geographic and operational linkages between firms, but also how business operations are impacted by ICTs, how telecommunications providers approach service provision with businesses and how telecommunication providers interact with each other.

Facet A: business to business interactions

This facet of the matrix considers the strength of the linkages between firms and the agglomerative benefits from co-locating near other firms in the same and related industries. The formation of such regional business clusters can improve business performance in several ways.[4] First, business clusters are often associated

Figure 6.1 The Broadband-Business Nexus.

with increased levels of regional social capital, particularly in urban and some suburban locales (Storper & Venables, 2004). This refers to the informational and relational benefits that accumulate amongst employees and area companies (Steinfield, Scupola, & López-Nicolás, 2010). In this context, Facet A is tele-communications agnostic. Connections are expected, but interaction takes place across a range of enabling technologies, face-to-face activities, local events or business transactions with a shared foci.

   Second, regional business clusters, including rural ones, encourage both innovation and competition between local industries, as well as enhancing the development of local resources (Porter, 2000; Breschi & Malerba, 2001; Steinfield, LaRose, Chew, & Tong, 2012). These local resources include, but are not limited to, broadband and related ICTs. The connection between broadband and small- and medium-sized firms is particularly crucial for rural locales because they often lack the transaction economies that exist in larger, more metropolitan regions (Leamer & Storper, 2001). In fact, recent research (Steinfield et al., 2012) suggests that in addition to regional clusters fueling business success, the importance that rural companies within such clusters place on broadband use in their business processes is also a significant predictor of business success.

Facet B: businesses to telecommunications interactions

Business operations are enhanced with access to broadband and related ICT services (Black & Lynch, 2001; Ford & Koutsky, 2005; Kummerow & Chan Lun, 2005). Information technology intensive sectors are especially tied to this, where broadband facilitates the ability to perform both simple and complex business tasks more efficiently. Again, *participation* uses are considered the most basic applications of ICT. These include email, web-browsing and electronic sharing of documents (Forman et al., 2005a). *Enhancement* uses of the Internet are those that fundamentally change business operations or allow companies to implement and/or offer new products and services (Forman et al., 2005a). In some cases, broadband telecommunications technologies may also be the "core" enabling technology of a business (e.g., finance, information technology and data warehousing). In fact, recent research on new information and communication technology (NICT) clusters has demonstrated that the broadband mediated interaction between vertically related industries has been beneficial for companies in their efforts to reduce barriers to knowledge exchange. This often includes online collaboration (May & Carter, 2001; Munkvold, 2003), which can dramatically enhance the efficiency of the product development process while simultaneously spurring economic growth (Krafft 2004, 2010; Liu, Dicken, & Yeung, 2004) for a region. But it is also important to note that broadband is beginning to facilitate more complex collaborative efforts such as the development of frameworks for open innovation (Schaffers et al., 2011), offshore research and development networks, and firms that leverage collective intelligence for optimizing global operations (Komninos, 2009).

Facet C: telecommunications to business interactions

Telecommunication providers often target business-rich locations to provide high-bandwidth services and/or bundled packages (voice and data) for small- and medium-sized enterprises (SMEs). There are several reasons for this. First, demand for residential broadband services is somewhat more elastic (Crandall, Sidak, & Singer, 2002; Rappoport, Kridel, Taylor, Duffy-Deno, & Allemen, 2003; Cardona, Schwarz, Yurtoglu, & Zulehner, 2009) than business demand. Second, small firms tend to be more sensitive to price than larger firms. Third, geographic clusters of information intensive businesses allow providers an opportunity for rolling out advanced platforms to build economies of scale and scope. These economies of scale mean broadband providers are willing to make infrastructure investments, such as upgrading the capacity of fiber trunks, central offices or local cable nodes with the knowledge that a return on such investments is likely.

Firms that are early adopters embrace these rollouts because they are seeking a competitive advantage. This means that broadband providers often see an immediate return on their investment and can work collaboratively with early adopters in the development of test and experimental platforms (TEP) that help work out the kinks with new technologies (Ballon, Pierson, & Delaere, 2007).

Other related strategies that benefit provides but that are not necessarily business-centric include the development of living labs as a strategy for promoting future service ecosystems over next generation networks (Pascu & van Lieshout, 2009). This allows for a variety of market and non-market parties to provide input and strategies for addressing potential barriers in the diffusion of new broadband technologies.

Facet D: telecommunications to telecommunications interactions

The co-location of telecommunication firms is critical for ensuring that high throughput systems such as business broadband and fiber continue to function both reliably and securely. Several factors are at play here. All broadband providers are reliant on viable peering and/or transit relationships to deliver content on the Internet. Although "peering" is a somewhat slippery term, it generally refers to a relationship established between two or more broadband providers for the purpose of exchanging traffic directly between networks.[5] This direct exchange of traffic at network access points (NAPs) avoids the use of third-party backbone networks and, in theory, minimizes costs. However, because maintaining a presence in a NAP is not free, many broadband providers have invested in private peering points where traffic is exchanged. Peering relationships are complicated and one needs to look no further than the recent troubles between Level 3 and Comcast (Yegulalp, 2014). In short, Comcast wants to charge Level 3 more money for delivering content from providers such as Netflix. Level 3 suggests that Comcast is throttling Netflix speeds because Level 3 is refusing to pay.[6] The important point to make here is that the presence of high-capacity peering and transit points, telecommunication hotels, data centers and redundant infrastructure for mission-critical applications is increasingly important for business continuity. Further, although the impact of these relationships is regional, super-regional or transcontinental, network operations dictate that interactions occur locally. As a result, the clustering of telecommunication resources has played an essential role in shaping the coverage and associated resilience of the underlying telecommunications infrastructure (Grubesic & O'Kelly, 2002; Grubesic & Murray, 2005; Weller & Woodcock, 2013). Geographical locations with multiple peering points, redundant power infrastructure and a concentration of diverse fiber backbones are attractive to data or information intensive industries. Not only does competition spur better QoS and more competitive pricing, but many industries (e.g., finance) rely on diverse and redundant connections to ensure operational continuity and the swift execution of internet-mediated transaction. Therefore, the presence of such critical infrastructure may be a differentiating location factor for many businesses.

A visual snapshot of the BBN

Although the various facets of the BBN were discussed individually and are represented discretely in Figure 6.1, there is a high level of complementarity between each component. One obvious issue worth addressing here is how the relationships displayed in Figure 6.1 might actually manifest in a region. Using FCC

Form 477 data for 2004, which detailed broadband provider counts by ZIP code and combining this with a 2004 business point database, aggregated to ZIP codes (Mack & Grubesic, 2014), both univariate and bivariate measures of spatial auto-correlation (Anselin, Syabri, & Kho, 2006) are applied to explore each facet of the BBN for the state of Ohio. This is not meant to be a comprehensive analysis, nor a detailed technical treatise of the spatial clustering of businesses and telecommunication providers. It is purely exploratory and is used to facilitate further discussion of the BBN.

Figure 6.2 highlights the geographic setting for this visual snapshot. The state of Ohio provides an interesting study area for this type of exploratory analysis. With a total population of approximately 11.5 million, approximately 22% of Ohio's residents live in rural areas (USDA, 2013). Many large urban complexes are also found in the state, including Cleveland, Columbus, Cincinnati, Dayton and Akron. Ohio's economy is relatively diverse, but strongly rooted in trade/transportation/utilities – employing nearly 19% of Ohio's workforce

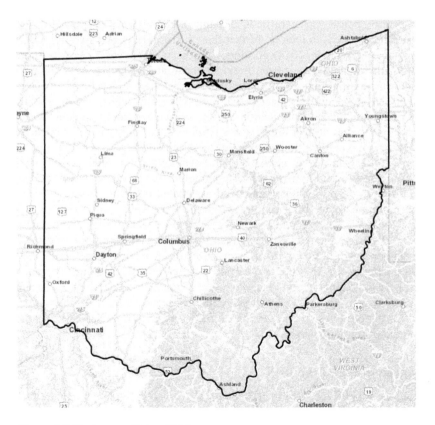

Figure 6.2 Study area: State of Ohio.

(ODOD, 2014). Ohio is also strong in health care and education (17%), professional and technical services (13%) and manufacturing (13%; ODOD, 2014). These relative shares of employment in each sector have remained relatively consistent during the past decade. Over 50 Fortune 1,000 firms are headquartered in Ohio, including Kroger (retail), Nationwide (insurance), Progressive (insurance), Proctor & Gamble (consumer goods), Fifth Third (banking) and Cardinal Health (healthcare).

Table 6.1 details the descriptive statistics for broadband providers and firm counts for each ZIP code area in the state of Ohio for 2004. There were 1,220 ZIP code areas in the state with an average of 220 firms and 4.6 broadband providers located in each. Table 6.1 also highlights average establishment and broadband provider counts for each of the four BBN facets. However, the most relevant portion of this exploratory analysis is the results displayed in Figures 6.3a–d, which highlight the spatial distribution of the four BBN facets for Ohio.

Figure 6.3a (business to business) was derived using a univariate measure of spatial autocorrelation on businesses, by ZIP code, for the entire state.[7]

Table 6.1 Average establishment and broadband provider counts by ZIP code area in Ohio, 2004

	Establishments	*Broadband providers*
Ohio ZIP code areas (n = 1220)	220.32	4.60
Facet A: Business to business		
High-High (n = 64)	741.95	10.29
Low-Low (n = 46)	50.32	2.95
Low-High (n = 49)	92.34	4.24
High-Low (n = 17)	678.11	7.35
Facet B: Business to telecommunication		
High-High (n = 121)	711.08	10.98
Low-Low (n = 37)	69.27	3.43
Low-High (n = 60)	115.13	6.40
High-Low (n = 9)	520.33	2.66
Facet C: Telecommunication to business		
High-High (n = 186)	663.42	9.93
Low-Low (n = 32)	25.59	1.75
Low-High (n = 22)	75.09	1.54
High-Low (n = 32)	409.25	6.56
Facet D: Telecommunication to telecommunications		
High-High (n = 211)	589.71	10.30
Low-Low (n = 23)	59.43	2.00
Low-High (n = 17)	35.58	1.23
High-Low (n = 24)	239.24	5.84

The local Moran's *I* value of 0.3324 ($p = 0.05$) suggests a moderate, yet statistically significant level of positive autocorrelation.[8] Specifically, Figure 6.3a indicates that businesses are clustering in the major metropolitan areas of Ohio (Cleveland, Columbus, Cincinnati, Dayton, Akron/Canton and Toledo). This result generally supports the theoretical tenets of the BBN detailed in the previous section. Namely, firms gravitate to larger urban and suburban areas, often forming business clusters. These clusters are typically fueled by regional social capital and local transaction economies. The remaining patterns for the state are somewhat uneven, but suggest a patchwork (low-low and high-low) activity throughout Southeastern Ohio (i.e., Appalachian Ohio), one of the most economically challenged and isolated regions of the state.

a) Business to business

b) Businesses and telecommunications

c) Telecommunications and businesses

d) Telecommunications and telecommunications

Not Significant
High-High
Low-Low
Low-High
High-Low

N

Figure 6.3 Spatial dimensions of the Broadband-Business Nexus.

Figure 6.3b (businesses and telecommunications) was derived using a bivariate measure of spatial autocorrelation. Business counts by ZIP code are lagged with broadband provision values; this measure of spatial association explores the degree to which the presence of more businesses is associated with higher levels of broadband provision. The Moran's I value of 0.3075 ($p = 0.05$), suggests that a moderate level of positive spatial autocorrelation exists between these entities. While the results must be interpreted cautiously, it is clear that businesses and broadband are co-located in Ohio, particularly in urban areas (e.g., high-high). The descriptive statistics in Table 6.1 confirm this. The core business areas (high-high) also have the highest average count of broadband providers for the state (10.98). However, there is some variation to this pattern. The major difference between Figures 6.3a and 6.3b can be found in the Columbus, Dayton and Cincinnati corridor, where there is a stronger spatial envelopment of the core areas (high-high) by low-high ZIP code areas. This highlights the spatial reach of broadband into the metropolitan periphery where robust residential markets exist, but it also suggests the lack of a strong business presence in these exurban locales. Appalachia continues to suffer from low levels of both business presence and telecommunications infrastructure, which is exemplified by the many low-low ZIP code areas in the region.

Figure 6.3c (telecommunications and businesses) reverses the bivariate structure and explores patterns associated with broadband provision levels lagged with business counts. This measure of spatial association explores the degree to which higher broadband provision levels are associated with higher business counts. The Moran's I value of 0.3437 ($p = 0.05$) suggests that the data are positively autocorrelated with the bulk of the high-high clusters located in large metropolitan areas. The main takeaway from this figure is that telecommunications providers likely target areas rich with businesses, albeit not exclusively. As detailed previously, there are many incentives for broadband providers to do this – inelastic demand from firms, opportunities for rolling out advanced platforms and building economies of scale and scope. As with the previous tests for autocorrelation, portions of Appalachian Ohio (southeast) struggle to attract the combination of businesses and broadband. However, many of the high-low ZIP code areas are found in this corner of Ohio, suggesting that broadband is available, but business counts are low. Does this debunk the theory that telecommunications technologies will facilitate decentralization? Not necessarily. It is important to remember that although ICTs function as a differentiating factor in business location decisions, they are not the only factor. That said, Figure 6.3c suggests that even when broadband is available in rural areas, there is no guarantee that businesses will gravitate to these regions.

Finally, Figure 6.3d (telecommunications and telecommunications) uses a univariate test of local spatial association to highlight broadband clusters within the state of Ohio. With a Moran's I value of 0.4214 ($p = 0.05$), the strongest spatial association found for the four BBN facets emerges. The majority of the high-high ZIP code areas remain in the largest metropolitan complexes (Cleveland, Akron/Canton, Columbus, Dayton, Cincinnati and Toledo). Although these core areas have the largest share of broadband providers for the state, there are other locations throughout the state that register in the high-low category, including many ZIP code areas in Appalachian Ohio. However, as detailed by Table 6.1, when compared to the core (high-high) regions, these locales have about 50% fewer broadband providers on average. Simply put, although the level of provision is higher in these areas when compared to their immediate neighbors, it is still far short of the average provider counts found within the major metropolitan areas of the state.

Discussion and conclusion

This preliminary spatial analysis of the BBN in the state of Ohio reveals important local nuances in each of its four dimensions. Given these findings, national level studies of the BBN over time can provide important information about the evolution and relative dynamism of the relationships between businesses and telecommunications provision. Greater resolution on positive feedbacks between broadband and businesses can yield valuable information about constructing local solutions for mitigating gaps in broadband access. In this regard, the framework outlined in this chapter highlights the potential for synergies between efforts to improve broadband provision levels and efforts to retain and attract local area businesses. These synergies may prove particularly important for exurban locales that have a cost advantage over expensive central city locations, but have historically had trouble in attracting broadband providers (Grubesic, 2008). In these instances, businesses can fill the demand gap that is critical to attracting this critical infrastructure.

That said, much more work remains for developing a comprehensive understanding of each facet in the BBN. While econometric studies are underway to unravel the broadband to business dimension of this nexus (Mack et al., 2011; Mack and Rey, 2014), the evolution of this dimension, over time, remains an understudied aspect of this relationship. While studies have found that broadband is an important location factor for businesses, particularly when compared to other types of infrastructure (airports, highways,) and workforce characteristics (Mack et al., 2011; Mack & Rey, 2014), there are reasons to suspect that the gradual diffusion of broadband networks may have diminished their value to firms over time. Alternatively, the tendency for providers to upgrade first-tier, densely populated locales with next generation technologies (Grubesic, 2015) may mean infrastructure quality inequities will endure over time. These are important nuances to acknowledge, especially if the research and policy communities are to better understand the impacts of infrastructure deployment efforts on business presence – which is perhaps one of the more important aspects of the competitive advantages endowed to regional economies by broadband.

Notes

1 We use the term "thickness" as a reference to the amount of co-located telecommunication infrastructure.

2 LATAs are basically service areas and are denoted with area codes (e.g., 215, 503). There are approximately 200 in the U.S.

3 By definition, right of way refers to a legal right of passage over land (or under it) owned by another. Telephone companies, CATV companies and cellular providers are required to obtain right-of-way to trench, place wireless antenna structures or plant poles (Newton, 2013).

4 Depending on regional structure, these clusters can adhere to the core facets of both the concentration and heterogeneous effects school of thought.

5 Transit refers to a relationship where one network operator (A) pays money (i.e., settlement) to another network operator (B) for the right to transmit traffic from network (A) across network (B).

6 There are certainly other factors at play here, including the somewhat murky arrangement for peering between both networks. For more details, see Yegulalp (2014).

7 A spatial weights matrix based on queen contiguity was used to generate Figures 6.3a–d. A total of 9,999 permutations were used to test for statistical significance.

8 High-high corresponds to ZIP code areas displaying high levels of businesses or broadband providers that are surrounded by other ZIP code areas with similar values. Low-low corresponds to ZIP code areas displaying low levels of businesses or broadband providers that are surrounded by other ZIP code areas with similar values. Low-high corresponds to ZIP code areas displaying low levels of businesses or broadband providers that are surrounded by ZIP codes displaying relatively higher values. Finally, high-low corresponds to ZIP code areas displaying relatively high levels of businesses or broadband providers that are surrounded by ZIP code areas displaying relatively lower values.

References

96 Act [Telecommunications Act of 1996]. (1996). Pub. LA. No. 104-104, 110 Stat. 56.

Abbate, J. (1999). *Inventing the Internet*. Cambridge: The MIT Press.

Andronico, G., Ardizzone, V., Barbera, R., Becker, B., Bruno, R., Calanducci, A. & Scardaci, D. (2011). e-Infrastructures for e-Science: A global view. *Journal of Grid Computing, 9*(2), 155–184.

Anselin, L., Syabri, I., & Kho, Y. (2006). GeoDa: An introduction to spatial data analysis. *Geographical Analysis, 38*(1), 5–22.

Arbore, A., & Ordanini, A. (2006). Broadband divide among SMEs: The role of size, location and outsourcing strategies. *International Small Business Journal, 24*(1), 83–99.

Atkinson, R.D. (1998). Technological change and cities. *Cityscape: A Journal of Policy Development and Research, 3*(3), 129–170.

Audirac, I. (2005). Information technology and urban form: Challenges to smart growth. *International Regional Science Review, 28*(2), 119–145.

Aufderheide, P. (1999). *Communications policy and the public interest: The Telecommunications Act of 1996*. New York: Guilford Press.

Aufderheide, P. (2006). 1996 Telecommunications Act: Ten years later. *Federal Communication Law Journal, 58*, 407.

Ballon, P., Pierson, J., & Delaere, S. (2007). Fostering innovation in networked communications: test and experimentation. Designing for networked communications. *Strategies and Development*, 137.

Black, S.E., & Lynch, L.M. (2001). How to compete: The impact of workplace practices and information technology on productivity. *Review of Economics and Statistics, 83*(3), 434–445.

Breschi, S., & Malerba, F. (2001). The geography of innovation and economic clustering: Some introductory notes. *Industrial and Corporate Change*, *10*, 817–834.

Brogan, P. (2013). U.S. broadband provider investment growing. *U.S. Telecom*. Retrieved from http://tinyurl.com/poe2tcc

Cairncross, F. (1997). *The death of distance*. Boston: Harvard Business School Press.

Cambini, C., & Jiang, Y. (2009). Broadband investment and regulation: A literature review. *Telecommunications Policy*, *33*(10), 559–574.

Capello, R. (1994). Towards new industrial and spatial systems: the role of new technologies. *Papers in Regional Science*, *73*(2), 189–208.

Capello, R., & Nijkamp, P. (1996). Telecommunications technologies and regional development: Theoretical considerations and empirical evidence. *Annals of Regional Science*, *30*(1), 7–30.

Cardona, M., Schwarz, A., Yurtoglu, B.B., & Zulehner, C. (2009). Demand estimation and market definition for broadband Internet services. *Journal of Regulatory Economics*, *35*(1), 70–95.

Center for an Urban Future. (2004). *New York's broadband gap*. Retrieved from https://nycfuture.org/research/publications/new-yorks-broadband-gap_

Chakraborty, J., & Bosman, M. (2005). Measuring the digital divide in the United States: Race, income, and personal computer ownership. *The Professional Geographer*, *57*, 395–410.

Crandall, R.W. (2005). *Competition and chaos: US telecommunications since the 1996 Telecom Act*. Washington, DC: Brookings Institution Press.

Crandall, R.W., Hahn, R.W., Litan, R.E., & Wallsten, S. (2004). Bandwidth for the people. *Policy Review*, (127), 04–01.

Crandall, R.W., Sidak, J.G., & Singer, H.J. (2002). The empirical case against asymmetric regulation of broadband Internet access. *Berkeley Law and Technology Journal*, *17*(1), 953–987.

Czernich, N., Falck, O., Kretschmer, T., & Woessmann, L. (2011). Broadband infrastructure and economic growth. *The Economic Journal*, *121*(552), 505–532.

District of Columbia Circuit Court [DCCC]. (2004). United States Telecom Association v. Federal Communications Commission and United States of America (Respondents) and Bell Atlantic Telephone Companies (Intervenors). Retrieved from http://tinyurl.com/ott2eqq_

Downes, T., & Greenstein, S. (2007). Understanding why universal service obligations may be unnecessary: The private development of local Internet access markets. *Journal of Urban Economics*, *62*(1), 2–26.

Duffy-Deno, K.T. (2003). Business demand for broadband access capacity. *Journal of Regulatory Economics*, *24*(3), 359–372.

Ford, G.S., & Koutsky, T.M. (2005). Broadband and economic development: A municipal case study from Florida. *Review of Urban & Regional Development Studies*, *17*(3), 216–229.

Forman, C. (2005). The corporate digital divide: Determinants of Internet adoption. *Management Science*, *51*(4), 641–654.

Forman, C., Goldfarb, A. & Greenstein, S. (2003). The geographic dispersion of commercial Internet use. In S. Wildman and L. Cranor *(Eds.), Rethinking rights and regulations: Institutional responses to new communication technologies (*pp. 113–145). Cambridge: MIT Press.

Forman, C., Goldfarb, A. and S. Greenstein. (2005a). How did location affect adoption of the commercial Internet? Global village vs. urban leadership. *Journal of Urban Economics*, *58*(3), 389–420.

Forman, C., Goldfarb, A. & Greenstein, S. (2005b). How do industry features influence the role of location on Internet adoption? *Journal of the Association for Information Systems*, 6(12), 383–408.

Frieden, R. (2005). Lessons from broadband development in Canada, Japan, Korea and the United States. *Telecommunications Policy*, 29(8), 595–613.

Gabel, D., & Kwan, F. (2001). Accessibility of broadband telecommunication services by various segments of the American population. In Compaine, B.M. and Greenstein, S. (Eds.), *Communications policy in transition: The Internet and beyond* (pp. 295–320). Cambridge: MIT Press.

García, P.P., Thapa, B., & Niehaves, B. (2014). Bridging the digital divide at the regional level? The effect of regional and national policies on broadband access in Europe's regions. In *Electronic government* (pp. 218–229). Berlin Heidelberg: Springer.

Gaspar, J., & Glaeser, E.L. (1998). Information technology and the future of cities. *Journal of Urban Economics*, 43(1), 136–156.

Gibbons, J., & Ruth, S. (2006). Municipal Wi-Fi: Big wave or wipeout? *IEEE Internet Computing*, 10(3), 66–71.

Gibbs, D. (2001). Harnessing the information society? European Union policy and information and communication technologies. *European Urban and Regional Studies*, 8(1), 73–84.

Gibbs, D., & Tanner, K. (1997). Information and communication technologies and local economic development policies: The British case. *Regional Studies*, 31, 765–774.

Gillespie, A., & Williams, H. (1988). Telecommunications and the reconstruction of regional comparative advantage. *Environment and Planning A*, 20(10), 1311–1321.

Gillett, S.E., Lehr, W.H., & Osorio, C. (2004). Local government broadband initiatives. *Telecommunications Policy*, 28(7), 537–558.

Goddard, J.B., & Gillespie, A.E. (1986). Advanced telecommunications and regional economic development. *Geographical Journal*, 152(3), 383–397.

Google. (2014). *Google fiber*. Retrieved from https://fiber.storage.googleapis.com/legal/googlefibercitychecklist2-24-14.pdf

Graham, S., & Marvin, S. (2001). *Splintering urbanism: Networked infrastructures, technological mobilities and the urban condition*. London: Routledge.

Greenstein, S. (2007). *Data constraints & the internet economy: Impressions & imprecision*. NSF/OECDmeeting on "Factors Shaping the Future of the Internet." Retrieved from http://www.oecd.org/dataoecd/5/54/38151520.pdf

Grubesic, T.H. (2003). Inequities in the broadband revolution. *Annals of Regional Science*, 37, 263–289.

Grubesic, T.H. (2004). The geodemographic correlates of broadband access and availability in the United States. *Telematics and Informatics*, 21(4), 335–358.

Grubesic, T.H. (2006). A spatial taxonomy of broadband regions in the United States. *Information Economics and Policy*, 18, 423–448.

Grubesic, T.H. (2008). The spatial distribution of broadband providers in the United States: 1999–2004. *Telecommunications Policy*, 32(3), 212–233.

Grubesic, T.H. (2012). The US national broadband map: Data limitations and implications. *Telecommunications Policy*, 36(2), 113–126.

Grubesic, T.H. (2015). The broadband provision tensor. *Growth and Change*, 46(1), 58–80.

Grubesic, T.H., & Horner, M.W. (2006). Deconstructing the divide: Extending broadband xDSL services to the periphery. *Environment and Planning B*, 33(5), 685–704.

Grubesic, T.H., & Murray, A.T. (2002). Constructing the divide: Spatial disparities in broadband access. *Papers in Regional Science*, 81(2), 197–221.

Grubesic, T.H., & Murray, A.T. (2004). Waiting for broadband: Local competition and the spatial distribution of advanced telecommunication services in the United States. *Growth and Change, 35*(2), 139–165.

Grubesic, T.H. and Murray, A.T. (2005). Spatial-historical landscapes of telecommunication network survivability. *Telecommunications Policy, 29*(11), 801–820.

Grubesic, T.H., & O'Kelly, M.E. (2002). Using points of presence to measure city accessibility to the commercial Internet. *Professional Geographer, 54*(2), 259–278.

Grubesic, T.H., Matisziw, T.C., & Murray, A. (2011). Market coverage and service quality in digital subscriber lines infrastructure planning. *International Regional Science Review. 34*(3), 368–390.

Heger, D., Veith, T., & Rinawi, M. (2011). The effect of broadband infrastructure on entrepreneurial activities: The case of Germany. ZEW-Centre for European Economic Research Discussion Paper, (11-081).

Hoffman, D.L. & Novak, T. (1999). The evolution of the digital divide: Examining the relationship of race to Internet access and usage over time. Retrieved from http://dl.acm.org/citation.cfm?id=762630

Horvath, L. (2001). Collaboration: the key to value creation in supply chain management. *Supply Chain Management: An International Journal, 6*(5), 205–207.

Jackson, K.T. (1985). *Crabgrass frontier: The suburbanization of the United States.* Oxford University Press.

Jorgenson, D.W. (2001). Information technology and the U.S. economy. *American Economic Review, 91*, 1–32.

Kandilov, I.T. & Renkow, M. (2010). Infrastructure investment and rural economic development: An evaluation of USDA's broadband loan program. *Growth and Change, 41*(2), 165–191.

Karshenas, M., & Stoneman, P.L. (1993). Rank, stock, order, and epidemic effects in the diffusion of new process technologies: An empirical model. *The RAND Journal of Economics, 24*(4), 503–528.

Kim, J., & Gerber, B. (2005). Bureaucratic leverage over policy choice: Explaining the dynamics of state-level reforms in telecommunications regulation. *Policy Studies Journal, 33*(4), 613–633.

Kloch, C., Petersen, E.B., & Madsen, O.B. (2011). Cloud based infrastructure, the new business possibilities and barriers. *Wireless Personal Communications, 58*(1), 17–30.

Kolko, J. (2010). How broadband changes online and offline behaviors. *Information Economics and Policy, 22*, 144–152.

Komninos, N. (2009). Intelligent cities: Towards interactive and global innovation environments. *International Journal of Innovation and Regional Development, 1*(4), 337–355.

Korek, M., & Olszewski, R. (1981). Telecom – The winds of change. *Datamation, 27*(5), 160.

Kraemer, K.L. (1982). Telecommunications/transportation substitution and energy conservation: Part 1. *Telecommunications Policy, 6*(1), 39–59.

Krafft, J. (2004). Entry, exit and knowledge: Evidence from a cluster in the info-communications industry. *Research Policy, 33*(10), 1687–1706.

Krafft, J. (2010). Profiting in the info-coms industry in the age of broadband: Lessons and new considerations. *Technological Forecasting & Social Change, 77*, 265–278.

Kummerow, M., & Chan Lun, J. (2005). Information and communication technology in the real estate industry: Productivity, industry structure and market efficiency. *Telecommunications Policy, 29*(2), 173–190.

Kutay, A. (1986). Optimum office location and the comparative statics of information economies. *Regional Studies, 20*(6), 551–564.

Kutay, A. (1988a). Technological change and spatial transformation in an information economy: 1. A structural model of transition in the urban system. *Environment and Planning A, 20*(5), 569–593.

Kutay, A. (1988b). Technological change and spatial transformation in an information economy: 2. The influence of new information technology on the urban system. *Environment and Planning A, 20*(6), 707–718.

LaRose, R., Gregg, J.L., Strover, S., Straubhaar, J. & Carpenter, S. (2007). Closing the rural broadband gap: Promoting adoption of the Internet in rural America. *Telecommunications Policy, 31*, 359–373.

Leamer, E.E. & Storper, M. (2001). The economic geography of the Internet age. *Journal of International Business Studies, 32*(4), 641–665.

Lehr, W., Sirbu, M., & Gillett, S. (2004). Municipal wireless broadband: Policy and business implications of emerging access technologies. Paper prepared for "Competition in Networking: Wireless and Wireline," London Business School.

Liu, W., Dicken, P., & Yeung, H.W. (2004). New information and communication technologies and local clustering of firms: A case study of the Xingwang Industrial Park in Beijing. *Urban Geography, 25*(4), 390–407.

Mack, E.A. (2014). Broadband and knowledge intensive firm flusters: Essential link or auxiliary connection? *Papers in Regional Science, 93*(1), 3–29.

Mack, E.A., Anselin, L., & Grubesic, T.H. (2011). The importance of broadband provision to knowledge intensive firm location. *Regional Science Policy & Practice, 3*(1), 17–35.

Mack, E.A., & Faggian, A. (2013). Productivity and broadband: The human factor. *International Regional Science Review, 36*(3), 392–423.

Mack, E., & Grubesic, T. (2009). Broadband provision and firm location in Ohio: An exploratory spatial analysis. *Tijdschrift voor Economische en Sociale Geografie, 100*(3), 298–315.

Mack, E.A., & Grubesic, T.H. (2014). U.S. broadband policy and the spatio-temporal evolution of broadband markets. *Regional Science, Policy, and Practice, 6*(3), 291–308.

Mack, E.A., & Rey, S.J. (2014). An econometric approach for evaluating the linkages between broadband and knowledge intensive firms. *Telecommunications Policy, 38*(1), 105–118.

Martínez, D., Rodriguez, J., & Torres, J.L. (2010). ICT-specific technological change and productivity growth in the US: 1980-2004. *Information Economics and Policy, 22*, 121–129.

May, A., & Carter, C. (2001). A case study of virtual team working in the European automotive industry. *International Journal of Industrial Ergonomics, 27*(3), 171–186.

McCann, P. & Shefer, D. (2004). Location, agglomeration and infrastructure. *Papers in Regional Science, 83*(1), 177–196.

McCann, P. & Sheppard, S. (2003). The rise, fall and rise again of industrial location theory. *Regional Studies, 37*(6–7), 649–663.

Mieszkowski, P., & Mills, E.S. (1993). The causes of metropolitan suburbanization. *The Journal of Economic Perspective*s, 7(3), 135–147.

Mitcsenkov, A., Paksy, G., & Cinkler, T. (2011). Geography- and infrastructure-aware topology design methodology for broadband access networks (FTTx). *Photonic Network Communications, 21*(3), 253–266.

Mokhtarian, P.L. (1990). A typology of relationships between telecommunications and transportation. *Transportation Research Part A: General, 24*(3), 231–242.

Moss, M.L. (1998). Technology and cities. *Cityscape: A Journal of Policy Development and Research, 3*(3), 107–127.

Mossberger, K., Tolbert, C.J., & McNeal, R.S. (2006). *Digital citizenship: The Internet, society, and participation.* Cambridge: MIT Press.

Mudambi, R. (2008). Location, control and innovation in knowledge-intensive industries. *Journal of Economic Geography, 8*(5), 699–725.

Munkvold, B.E. (Ed.). (2003). *Implementing collaboration technologies in industry: Case examples and lessons learned.* Springer.

Murray, A.T., & Grubesic, T. (Eds.). (2007). *Critical infrastructure: Reliability and vulnerability.* Springer Science & Business Media.

National Telecommunications and Information Administration [NTIA]. (2003). *Rights-of-way laws by state.* Retrieved from http://tinyurl.com/m5hujh8

Negroponte, N. (1995). *Being digital.* New York: Alfred A. Knopf.

Newton, H. (2013). *Newton's telecom dictionary.* CMP Books.

Nishida, T., Pick, J.B., & Sarkar, A. (2014). Japan's prefectural digital divide: A multivariate and spatial analysis. *Telecommunications Policy, 38*(11), 992–1010.

Norris, P. (2001). *Digital divide: Civic engagement, information poverty, and the Internet worldwide.* Cambridge, UK: Cambridge University Press.

Nucciarelli, A., Castaldo, A., Conte, E., & Sadowski, B. (2013). Unlocking the potential of Italian broadband: Case studies and policy lessons. *Telecommunications Policy, 37*(10), 955–969.

Ohio Department of Development [ODOD]. (2014). *Economic overview of the State of Ohio.* Retrieved from http://development.ohio.gov/files/research/E1000.pdf

O'Kelly, M.E., & Grubesic, T.H. (2002). Backbone topology, access, and the commercial Internet, 1997–2000. *Environment and Planning B, 29*(4), 533–552.

Parajuli, J. & Haynes, K.E. (2012). *Broadband Internet and new firm formation: A U.S. perspective.* George Mason School of Public Policy Research Paper No. 2013-03.

Pascu, C., & van Lieshout, M. (2009). User-led, citizen innovation at the interface of services. *Info, 11*(6), 82–96.

Pons-Novell, J., & Viladecans-Marsal, E. (2006). Cities and the internet: The end of distance? *Journal of Urban Technology, 13*(1), 109–132.

Porter, M.E. (2000). Location, competition, and economic development: Local clusters in a global economy. *Economic Development Quarterly, 14,* 15–34.

Prieger, J.E. (2013). The broadband digital divide and the economic benefits of mobile broadband for rural areas. *Telecommunications Policy, 37*(6), 483–502.

Prieger, J.E., & Hu, W. M. (2008). The broadband digital divide and the nexus of race, competition, and quality. *Information Economics and Policy, 20*(2), 150–167.

Ramsay, B. (2013). Catalysing regional business development through high speed broadband: Opportunities and risks. In Kinnear, S., Charters, K., and Vitartas, P. (Eds.), *Regional advantage and innovation* (pp. 269–287). Physica-Verlag HD.

Rappoport, P., Kridel, D., Taylor, L., Duffy-Deno, K., & Allemen, J. (2003). Residential demand for access to the Internet, Chapter5. In G. Madden (Ed.), *International handbook of telecommunications economics* (Vol. II). Cheltenham, UK: Edward Elgar.

Riddlesden, D., & Singleton, A.D. (2014). Broadband speed equity: A new digital divide? *Applied Geography, 52,* 25–33.

Röller, L.H., & Waverman, L. (2001). Telecommunications infrastructure and economic development: A simultaneous approach. *American Economic Review, 9*(4), 909–923.

Salomon, I. (1996). Telecommunications, cities and technological opportunism. *Annals of Regional Science, 30*(1), 75–90.

Schaffers, H., Komninos, N., Pallot, M., Trousse, B., Nilsson, M., & Oliveira, A. (2011). *Smart cities and the future internet: Towards cooperation frameworks for open innovation* (pp. 431–446). Berlin Heidelberg: Springer.

Shiels, H., McIvor, R., & O'Reilly, D. (2003). Understanding the implications of ICT adoption: Insights from SMEs. *Logistics Information Management, 16*(5), 312–326.

Sirbu, M., Lehr, W., & Gillett, S. (2006). Evolving wireless access technologies for municipal broadband. *Government Information Quarterly, 23*(3), 480–502.

Sohn, J., Kim, T.J. & Hewings, G.J.D. (2003). Information technology and urban spatial structure: A comparative analysis of the Chicago and Seoul regions. *Annals of Regional Science, 37*(3), 447–462.

Steinfield, C. (2004). Situated electronic commerce: A view as complement rather than substitute for offline commerce. *Urban Geography, 25*(4), 353–371.

Steinfield, C., Scupola, A., & López-Nicolás, C. (2010). Social capital, ICT use and company performance: Findings from the Medicon Valley Biotech Cluster. *Technological Forecasting and Social Change, 77*(7), 1156–1166.

Steinfield, C., LaRose, R., Chew, H.E., & Tong, S.T. (2012). Small and medium-sized enterprises in rural business clusters: The relation between ICT adoption and benefits derived from cluster membership. *The Information Society, 28*(2), 110–120.

Storper, M., & Venables, A.J. (2004). Buzz: Face-to-face contact and the urban economy. *Journal of Economic Geography, 4*(4), 351–370.

Townsend, L., Sathiaseelan, A., Fairhurst, G., & Wallace, C. (2013). Enhanced broadband access as a solution to the social and economic problems of the rural digital divide. *Local Economy, 28*(6), 580–595.

USDA (2013). *Data for rural analysis.* Retrieved from http://tinyurl.com/nh8mpup

Van Deursen, A.J., & Van Dijk, J.A. (2014). The digital divide shifts to differences in usage. *New Media & Society, 16*(3), 507–526.

Wallsten, S.J. (2005). *Broadband penetration: An empirical analysis of state and federal policies.* Retrieved from http://tinyurl.com/phtxl9h

Warf, B. (1989). Telecommunications and the globalization of financial services. *Professional Geographer, 41*(3), 257–271.

Warf, B. (1995). Telecommunications and the changing geographies of knowledge transmission in the late 20th century. *Urban Studies, 32*(2), 361–378.

Warf, B. (2003). Mergers and acquisitions in the telecommunications industry. *Growth and Change, 34*(3), 321–344.

Weller, D., & Woodcock, B. (2013). Bandwidth bottleneck [Data Flow]. *IEEE Spectrum Magazine, 50*(1), 80–80.

Whitacre, B., Gallardo, R., & Strover, S. (2014). Broadband's contribution to economic growth in rural areas: Moving towards a causal relationship. *Telecommunications Policy, 38*(11), 1011–1023.

Wright, D.J. (1992). Strategic impact of broadband telecommunications in insurance, publishing, and health care. *IEEE Journal on Selected Areas in Communications, 10*(9), 1369–1381.

Yegulalp, S. (2014). Level 3 accuses Comcast, other ISPs of 'deliberately harming' broadband service. *InfoWorld.* Retrieved from http://tinyurl.com/moyuxyc

Youtie, J. (2000). Field of dreams revisited: Economic development and telecommunications in LaGrange, Georgia. *Economic Development Quarterly, 14*(2), 146–153.

7 Digital innovation and entrepreneurship

As previous chapters highlighted, a large body of work has been dedicated to understanding both the human and spatial dimensions of the digital divide. Chapter 6, which deals with the Broadband-Business Nexus (BBN), explores the contextual and geospatial associations between broadband provision and business presence. The relationship between businesses, broadband and the many actors that drive these connections is extremely complex. In particular, there is growing interest in deepening our understanding of the association between broadband and entrepreneurial activity.

There are various ways of defining entrepreneurship. While scholars may not agree on a single definition, entrepreneurship may be defined as the process of creating a market and economic value for a particular innovation. In this context, an entrepreneur is different from a small business owner (Leigh & Blakely, 2013). The key distinction is that entrepreneurs create *new markets* and *new business opportunities* from an innovative idea, product, service or organizational structure or strategy, while small business owners have a business in a *predefined market* that deals with *existing* ideas, products, services and organizational structures or strategies (ibid.). While the distinction between entrepreneurial ventures and small businesses is difficult at times to ascertain, it is nevertheless important to make this distinction. An example of an entrepreneurial venture is eBay, which was one of the first companies to bring auctions and products sales to the online arena. A dry cleaner or clothing shop owner is an example of a small business. In the context of this study, we will use a surrogate for entrepreneurial activity, new firm formation. Although this will certainly include some small business owners, it will also include entrepreneurs.

Currently, it is presumed that broadband availability is an important ingredient to the business environment for entrepreneurs (Malecki, 1994; Felsenstein & Fleischer, 2002; Leigh & Blakely, 2013), but little empirical verification of this anticipated linkage exists. Work attempting to unravel this relationship at the state level has found that both geography and industrial composition are important moderating factors in the relationship between broadband and new business activity (Parajuli & Haynes, 2012). This is not an unexpected finding, but additional research is required. This work must include more fine-grained data given the geographic nuances to broadband provision discussed in Chapter 5.

Usage studies are also important because they can provide valuable evidence regarding adoption rates and insight into the ways that entrepreneurs use broadband to achieve cost and efficiency gains. Lack of work to date in the realm of entrepreneurship is a significant gap in knowledge, effectively creating a "blind spot" in broadband policy for the United States. Without this information, and additional feedback on the linkages between broadband and entrepreneurial activity, it is nearly impossible to evaluate the impacts of efforts associated with the National Broadband Plan (FCC, 2010) or the multi-billion dollar infrastructure investment/deployment rollouts by private corporations and the U.S. federal government (NTIA, 2014).

Given these issues, the purpose of this chapter is to provide a high-level overview of the linkages between technological change, broadband and entrepreneurial activity. This includes a theoretical rationale for these linkages, a discussion of broadband-enabled innovations and details regarding the importance of broadband as a general-purpose technology (GPT). A conceptual framework for understanding broadband as an enabling infrastructure for entrepreneurial activity is also developed. This framework will be employed to analyze a sub-dimension of the BBN, the *Broadband-Entrepreneurship Nexus* (BEN). In particular, the relationship between broadband and new business activity will be analyzed in both space and time, at the inter- and intra-metropolitan scales. This analysis highlights the utility of the conceptual framework developed in this chapter as well as important directions for future research.

General-purpose technologies (GPTs)

Before exploring the context of broadband as a GPT, it is important to provide a detailed background on GPTs and their economic importance. Put simply, GPTs are important because they are associated with significant economic transformations, including the industrial revolution and the information age (Bresnahan, 2010). GPTs are impactful because they are frequently used as inputs into a large and diverse array of industries (Helpman & Trajtenberg, 1994). That said, there can be delays between the time a GPT is introduced and the time when positive growth and associated economic benefits are realized. This is because the initial spread of new GPTs can be slow, as early adopters make investments in complementary areas and the laggards keep producing with and/or using older technologies. Thus, the return on investment does not occur immediately (Helpman & Trajtenberg, 1996). In short, there is a period of initial diffusion for the GPT, which is followed by a period of sustained growth, but this only happens after a sufficient number of sectors have adopted the GPT (ibid.).

During the initial phase of GPT diffusion, many early adopters and some laggards make additional investments in complementary products, such as printers and scanners to complement the personal computer, that pay off right away (ibid.). Ironically, these immediate returns on investment are most commonly attributed to GPTs and the period of initial adoption. Forgotten, is the learning that is required to fully embrace the new GPT as part of the business regime. In fact, this is exactly what happened with personal computers in the 1980s. Many scholars highlighted the presence of a productivity paradox during in the 1980s, which happened to correspond with the widespread diffusion of computer technologies through business and industry (Roach, 1987; Bailey & Gordon, 1988; Strausman 1990; Morrison, 1997).

Helpman and Trajtenberg (1994) coined two helpful phrases to disentangle these distinct phases of GPT introduction and diffusion, a "time to sow" and a "time to reap." The "time to sow" requires the use of resources to develop complementary inputs that will allow people to use the GPT. The "time to reap" is when the GPT becomes widely usable because enough complementary products have been developed (ibid.).

Economic impacts of GPTs

Jovanovic and Rousseau (2005) noted six specific economic symptoms associated with GPTs. These symptoms are associated with the "time to sow," when initial investments and learning produce a slowdown in economic activity, as people and businesses familiarize themselves with the new technology. The first symptom is a temporary slowdown in productivity as firms make investments and learn about the new technology. The second symptom is an increase in the skill premium for workers, which means that the earnings of highly skilled workers rise more quickly than the earnings of less-skilled workers. The third symptom is increased firm churn in the form of entry, exit and mergers. This symptom results in the acceleration of firm obsolescence (assuming no innovation), given the increased pace of technological change. This can also decrease the usefulness of firm physical capital (Jovanovic & Rousseau, 2002). The fourth symptom of GPTs is an initial fall in firm stock prices because of the decline in value of the physical capital that becomes outdated as a result of the arrival of the GPT (Jovanovic & Rousseau, 2005). The fifth symptom is realized advantage of newer and smaller firms in bringing the products and services associated with the GPT to market. This is likely because of the increased flexibility of firms that do not suffer from issues associated with technology and process lock-in that older firms face. The sixth and final symptom associated with the introduction of a GPT is a rise in interest rates and trade deficits.

Adding to the overall complexity of GPTs and their associated impacts is the strong temporal component associated with technological change (Murphy, Riddle, & Romer, 1998; Lipsey, Carlaw, & Bekar, 2005). Consider, for example, the problems associated with income disparities here in the U.S. In the first seven decades of the twentieth century, technological change helped close income disparities (Lipsey et al., 2005). Yet, in the closing decades of the twentieth century, technological change has contributed to the widening of income disparities between individuals (ibid.). One potential explanation for this growing income gap is the difference in the demand for more highly skilled and educated workers – a consequence of technological change (Murphy et al., 1998). Often categorized as "skills biased" technological change, an increased emphasis on acquiring workers with computer skills that also have commensurate levels of education, has widened the income gap (Bartel & Lichtenburg, 1987; Helpman & Rangel, 1998). Further, this trend is also indicative of technology-skills complementarity, which means the education requirements for new technologies are higher than the education requirements for older technologies (Helpman & Rangel, 1998). This complementarity between education and technology induces people to stay in school longer (ibid.). Thus, the link between education and human capital, particularly as it relates to technology, is becoming increasingly important. This is especially true as developed economies transition from manufacturing to a more service-oriented composition (Lloyd & Clark, 2001). In fact, studies have noted that human capital is linked to technology adoption (Benhabib & Spiegel, 1994) and the productivity impacts of technology (Baldwin, Diverty, & Sabourin, 1995; Entorf & Kramarz 1998; Bartelsman & Doms 2000; Gretton, Gali, & Parham, 2004).

Characteristics and types of GPTs

Aside from temporal and human capital-related variations in GPT impacts, one must also consider differences in the technologies classified as GPTs. For example, Lipsey et al. (2005) highlighted differences among GPTs and suggested that they be divided into two categories: 1) Transforming GPTs and 2) Other GPTs. Although this typology is simple, the two categories are important because it is the transforming GPTs that have the largest impact on economic activity (ibid.). It is estimated that over the last ten thousand years, 24 transforming GPTs have been created (ibid.). Table 7.1 contains a list of these 24 technologies along with the era in which they were introduced, and the type of technological change they represent (product, process or service). It also categorizes GPTs into six types: materials, power, information and communications technologies (ICTs), tools, transportation and organization. There are three key takeaways from Table 7.1. First, it reveals that the incidence of GPTs has increased, particularly in the twentieth century (ibid.). Second, 15 of these GPTs are product based. Third, of the GPTs that are product based, one-third are transportation oriented.

Even with all of this diversity in GPTs, they do share some commonalities. These facets are important to highlight, particularly as we move toward a discussion of the Internet as a GPT and its role in entrepreneurial activity and innovation. First, GPTs are different from other, less transformative technologies because they are widely adopted. Second, transformative GPTs have the potential for continued improvement over time. Finally, GPTs enable innovation (Bresnahan & Trajtenberg, 1995; Jovanovic & Rousseau, 2005; Bresnahan, 2010). To these three defining characteristics, Lipsey et al. (2005) add a fourth characteristic; transformative GPTs are generic, but distinctly recognizable over their entire lifetime.

Of these four characteristics, perhaps the most important and widely recognized facet of GPTs is their capacity to generate innovational complementarities. In short, innovational complementarities are what happens when continuous improvement

Table 7.1 A historical look at important GPTs

	Technology	Era of Development	Type of GPT	Class of GPT
1	Plant domestication	9000–8000 BC	Process	Material
2	Animal domestication	8500–7500 BC	Process	Material/Power/ Transportation
3	Ore smelting	8000–7000 BC	Process	Material
4	Wheel	4000–3000 BC	Product	Tool
5	Writing	3400–3200 BC	Process	ICT
6	Bronze	2800 BC	Product	Material
7	Iron	1200 BC	Product	Material
8	Waterwheel	Early medieval period	Product	Power
9	Three-masted sailing ship	15th century	Product	Transportation
		16th century	Process	ICT
10	Printing	Late 18th-early 19th centuries	Product	Power
11	Steam engine	Late 18th-early 19th centuries	Product	Organization
12	Factory system	Mid 19th century	Product	Transportation
13	Railway	Mid 19th century	Product	Transportation
14	Iron steamship	Late 19th century	Product	Power
15	Internal combusions engine	Late 19th century	Product	Power
16	Electricity	20th century	Product	Transportation
17	Motor vehicle	20th century	Product	Transportation
18	Airplane			
19	Mass production	20th century	Organizational	Organization
20	Computer	20th century	Product	ICT
21	Lean production	20th century	Organizational	Organization
22	Internet	20th century	Product	ICT
23	Biotechnology	20th century	Process	Material
24	Nanotechnology	Early 21st century	Process	Material

Source: Modified from Lipsey, Carlaw, and Bekar (2005) p. 132.

and application innovation features of GPTs are combined (Bresnahan, 2010). Lipsey, et al. (2005) also suggest that when such complementarities are discussed three sub-characteristics are evident: 1) they alter the value of existing technologies, 2) they provide opportunities to alter existing technologies and 3) they create opportunities to develop new technologies.

The Internet as a GPT

Although GPTs have radical, long-term impacts on people and the economy, it is important to note that many GPTs are not necessarily revolutionary in their underlying technology. Instead, they are radical in their application (Lipsey et al., 2005).

The extent of GPT impacts is also related to how they interact with existing technologies (ibid.). In this regard, the Internet is an exemplary GPT in many ways. Not only does it embody the core facets of GPTs detailed above, but the Internet is also a more recent technology that follows a long line of previous technological advancements that have become classified as GPTs including electricity and computers. In this way, the Internet is a nested innovation (Greenstein & Prince, 2006) that exhibits path dependency based on previous innovations. The Internet and related applications also represent the convergence of computers and telecommunications technology (Helpman & Trajtenberg, 1994). In fact, early studies showed that internet adoption was strongly tied to the diffusion of the PC (Jimeniz & Greenstein, 1998).[1] Thus, the combined impacts of the invention of the Internet (Abbate, 1999) and prior GPTs have led people to refer to these impacts as the ICT revolution.

The start date of the ICT revolution is a matter of some debate. Some scholars identify the invention of Intel's 4004 microprocessor, a foundational technology for personal computers, as the watershed moment (Jovanovic & Rousseau, 2005). Others identify the emergence of the New Economy in the 1990s as the genesis of the ICT revolution. During this era, productivity impacts associated with parallel advancements in the computer and telecommunications began to become evident in total factor productivity (TFP) statistics (Lipsey et al., 2005). Drilling deeper, Lipsey et al. (2005) suggest that the ICT revolution may have had its beginning during the World War II era, coinciding with the development of the computer, transistors and semiconductors. These developments eventually led to the development of integrated circuits and microprocessors many years later. Irrespective of when the ICT revolution began, the falling costs of computing in the 1980s helped computers and software become an increasingly large share of physical capital stock throughout the 1990s (Jovanovic & Rousseau, 2005). In fact, the price of computer capital fell faster than did diesel- or electricity-powered capital (Jovanovic & Rousseau, 2002) during this period.

As the ICT revolution bloomed, many notable companies emerged, including Intel, Microsoft, Apple, Cisco, Amazon and Google (Jovanovic & Rousseau, 2005; Lindgardt, Reeves, Stalk, & Deimler, 2009; Shaughnessy, 2013). Combined, the market capitalization of these six companies exceeded $1.2 trillion in May 2014 (Forbes, 2014). Google and Apple are particularly interesting because they represent business model innovators, which are companies with distinctive internal organizational structures and unique external relationships with customers and suppliers (Gambardella & McGahan, 2010). These two companies also represent a paradigm shift in the commercialization of assets that started in the 1980s and became particularly evident in the 1990s (ibid.). In particular, Google and Apple are representative of firms involved in intermediate technology markets, selling and/or licensing their ideas to companies that have more expertise in design, fabrication, commercialization, marketing and retailing (Gambardella & McGahan, 2010).

Broadband-related innovations

In many ways, the business model innovations detailed above are more strongly entrenched in companies that are tied to the Internet and, in turn, these innovations are reliant on the enabling capacity of high-speed broadband internet connections to fuel their development. However, these symbiotic innovations manifest in many different ways, adding to the mix of complementarities detailed above. Recently, the National Broadband Plan (FCC, 2010) highlighted three critical areas of innovation directly related to broadband: 1) networks, 2) devices and 3) applications. Since Chapter 3 covers the evolution of broadband infrastructure and its networks, the purpose of this section is to provide a brief overview of innovations in devices and applications. This is important because the downstream complementarities associated with the Internet and high-speed broadband connections helped spawn these innovations and presented entrepreneurs with a wide range of new business opportunities. As is discussed in the remaining portions of this chapter, both geography and regional context play an important role in digital innovation and entrepreneurial activity.

Device innovations

Device innovations enabled by the Internet have created an "Internet of things" (Ashton, 2009), or IoT, which is a generic term for the multitude of gadgets, appliances and other hardware connected to the Internet (Morgan, 2014). Broadband internet connections are largely responsible for aiding the proliferation of these connected devices as well as a host of other trends including local wireless systems, such as Wi-Fi, declining costs in technology and rapidly increasing smartphone usage (ibid.). It is estimated that the number of devices connected to the Internet will more than double from 16 billion connected devices in 2014, to 40.9 billion connected devices in 2020 (Press, 2014). Sensor nodes and/or networks, are anticipated to be the main source of device growth (Perera, Zaslavsky, Christen, & Georgakopoulos, 2014), but this expansion will also include non-hub devices such as activity trackers, smart meters, thermostats, clothing, watches and security systems (Press, 2014). All of these devices collect massive amounts of information pertaining to lifestyle and location. In turn, they offer users a range of strategies for making better and/or more efficient choices for work, recreation, travel and diet (Kelly, 2014). However, many challenges remain for making this ecosystem of devices both ubiquitous and truly seamless. This includes the basics, such as getting smart devices to communicate with each other across platforms (ibid.).

In 2015, smartphones remain the dominant innovation/device that uses broadband connections for portions of its functionality. Aside from enabling phone calls,[2] browsing the web or accessing social media, these devices frequently include capabilities for email, multimedia messaging service, music, navigation and photography. The first smartphone was IBM's Simon Personal Communicator, which was introduced in 1993 at a cost of $899 and required a two-year service agreement with BellSouth (Grush, 2012). Simon was available in 15 states and only 50,000 units were sold (Martin, 2014). Within the next few years, the Nokia 9000 was released – a device that included email, fax, web browsing and electronic organizing capacity (Businessweek, 1996). By 1997, after Ericsson released Penelope (which included a touch screen and stylus), the word "smartphone" was finally coined (Martin, 2014).

Over the next ten years, a variety of smartphone models were released by companies such as Palm, Nokia, Ericsson, Motorola and Samsung and usage surged between 2004 and 2007 (Martin, 2014). However, in January 2007, the smartphone landscape changed forever with the release of the Apple iPhone (McCarty, 2011), which spawned an entirely new generation of high-powered smartphones. In addition to this new breed of devices having significant economic, cultural and public health implications (Rogers, 2009; Laugesen & Yuan, 2010; Xing & Detert, 2010; Bever, 2014), perhaps the most important and relevant statistic is that 58% of U.S adults had a smartphone in January 2014. Although men held a slight edge in ownership (61%) when compared to women (57%) in the U.S. (Pew Research Center, 2014), it is projected that there will be close to 2.16 billion smartphone users worldwide by 2016, with the majority found in China, India and the United States (Emarketer, 2014).

Application innovations

Space constraints limit us from discussing the many application innovations that have emerged over the past several decades. However, there are several key innovations worth highlighting. First, perhaps the most critical software advance in internet history was the invention of the web browser. In the early years of the Internet, it was extremely difficult to locate and retrieve information (Abbate, 1999). Prior to web browsers, the Internet was primarily used and/or accessed by the scientific community (Moschovitis, Poole, & Senft, 1999). After Berners Lee proposed the idea for the World Wide Web (WWW) and several of its associated technologies (e.g., hypertext markup language), the production and consumption of data on the WWW exploded (Lamm, Reed, & Scullin, 1996; Berners Lee, Fischetti, & Dertouzos, 2000). This opened the door for e-commerce, internet broadcasting and a slew of other innovations (Berners Lee et al., 2000; Zook, 2003; Markoff, 2005).

A second major innovation was the development and release of the Java programming language by Sun Microsystems in 1995 (Moschovitis et al., 1999). The intrinsic value of Java was that application developers could write the Java code a single time and it could and/or would operate on all platforms that support Java without the need for modifying or recompiling the code (Oracle, 2015). This element of portability is critically important for the functionality of devices that belong to the larger broadband ecosystem. In particular, the use of Javascript allowed applications and/or programs to operate on web browsers regardless of the operating system or its vintage, as long as they supported Java. Today, Java is found in smart televisions, Blu-Ray players and Android-based smartphones.[3]

Although Apple products avoid the use of Java (Hess, 2013), it remains an important and popular development platform.

Third, it is essential to acknowledge the general trend of application development for smart devices. Consider, for example, the Apple iTunes App Store. In July 2014, it contained 1.2 million unique applications and had processed 75 billion downloads (Perez, 2014; Statista, 2014). During the same period, the Google Play Store hosted 1.3 million unique applications. This is an explosive level of growth in application development and consumer use considering neither of these stores existed prior to 2003. These user-oriented applications are not only critical to the larger internet ecosystem because of their popularity, but they also represent a new orientation to web content that is increasingly interactive and user-defined. An example of this is the advent of location-based services (LBS), which use real-time geographic data to provide information, services and security (Schiller & Voisard, 2004). Examples of LBS include Yelp, Foursquare and Google Maps (Goodrich, 2013). Push marketing notifications that make subscribers aware of sales or discounts at nearby restaurants and retailers are another example of LBS (Aalto, Göthlin, Korhonen, & Ojala, 2004).

Finally, in an effort to structure a synthesis of the widely varied innovations in application development, it is important to mention the work of David and Wright (1999). They made three keen observations/predications about computer applications that have come to fruition in the Internet age. First, an increasing number of information technologies have become *purpose built* and *task-specific*. This applies to many applications found in the iTunes and Google Play stores. The rollout of these task-specific applications in other sectors is also important. For example, basic scanner technologies and their associated applications are key technologies for logistic firms and retail outlets (e.g., supermarkets). In short, scanners and their software were created to log data, a domain where a mass-produced personal computer or smartphone would be a poor fit (David & Wright, 1999). Second, the use of personal computers as network servers has enabled client-server data processing systems (ibid.). No longer constrained by mainframes, terminals and their associated network topologies, the client-server data processing systems allow for a more complete utilization of local area networks for information and resource sharing (ibid.). Finally, the development of internet technology has revolutionized data processing *within* and *between* networks, as well as enhanced collaborative potential between people. Again, the development of Java as a cross-platform development language helps here, but there are many other technologies that enable platform agnostic collaboration including electronic conferencing tools (e.g., Skype), collaborative management tools (e.g., Central Desktop), and many others.

Entrepreneurial activity at the regional level

The brief overview of the web-related device and application innovations above clearly highlights the downstream complementarities associated with the Internet and high-speed broadband connections. Admittedly, this overview only scratched the surface, considering the multitude of other innovations in smart medical devices, internet-enabled drones, online payment systems, social networking and streaming media. The important point to be made here is that all of these innovations present entrepreneurs with a wide range of new business opportunities. The dot-com boom in the late 1990s and the soaring stock prices of companies such as Facebook and Google are evidence of the wealth-generating potential of the entrepreneurial opportunities associated with the Internet. In fact, scholars have suggested that the growth of the World Wide Web "spawned an entrepreneurial paradise" (Moschovitis et al., 1999, 156).

Despite the importance of technological change and entrepreneurship to national prosperity in the global information economy, existing theoretical and applied research in macroeconomics and entrepreneurship treat technological change and entrepreneurship as unrelated components of economic growth (Wennekers & Thurik, 1999; Acs, Audretsch, Braunerhjelm, & Carlsson, 2004; Audretsch & Keilbach, 2008). For several decades, economic growth theorists have grappled with the task of devising a generalized theory that accurately captures the role of technological change as a driving force of economic growth (Solow, 1956; Arrow, 1962; Griliches, 1979; Lucas, 1988; Barro & Sala-I-Martin, 1992; Mankiw, Romer, & Weil, 1992; Young 1993a, 1993b). While endogenous growth theorists have made great strides in revising neoclassical models to account for the role of endogenous rather than exogenous technological change on growth (Romer, 1986, 1990; Lucas, 1988; Aghion & Howitt, 1992), much more work is required to explicitly account for the role of the entrepreneur and their impact on economic growth (Wennekers & Thurik, 1999; Acs et al., 2004; Audretsch & Keilbach, 2008).

Regional context and the characteristics of the regional environment play an important role in fostering entrepreneurial activity (Audretsch & Fritsch, 1994; Tödtling & Wanzenböck, 2003; Wagener & Sternberg, 2004). Regional characteristics that account for variations in entrepreneurial activity include: the unemployment rate (Storey, 1991; Audretsch & Fritsch, 1994), the educational attainment of the workforce (Acs & Armington, 2004; Glaeser & Kerr, 2009), the size of establishments in a region (Armington & Acs, 2002), the relative urbanity of a region (Guesnier, 1994), and a region's industrial mix (Campi, Blasco, & Marsal, 2004; Delgado, Porter, & Stern, 2010).

Interestingly, although telecommunications infrastructure and regional technological capability are linked to the regional accumulation of technical and business knowledge (Malecki, 1994), few studies (save Parajuli & Haynes, 2012) have evaluated the impact of telecommunications infrastructure on variations in entrepreneurial activity. There has also been a comparative lack of attention dedicated to the sources of temporal variation in the linkages between regional entrepreneurial activity and economic growth uncovered in prior studies (Audretsch & Fritsch, 2002). These gaps are interesting because the availability of broadband is a likely source of variation in entrepreneurial activity over both space and time.

Evaluating the interaction between technological change and entrepreneurial activity is complicated by the varied endowments of telecommunications infrastructure across regions. This is a complex relationship to unravel for three reasons. First, as detailed in Chapter 5 and elsewhere, broadband infrastructure is heterogeneously distributed at a variety of spatial scales including urban and rural areas (Strover, 2001; Grubesic & Murray, 2004), metropolitan areas of different sizes (Moss & Townsend, 2000; O'Kelly & Grubesic, 2002), and *within* city limits (Graham, 1999, 2002; Grubesic & Murray, 2002). In part, this is because profit-seeking companies are responsible for deploying and updating the infrastructure. Second, the nature and quality of broadband is not homogeneous within or between regions. Recent studies show that in addition to significant variation in broadband platforms, prices and speeds also vary dramatically within and between regions (Grubesic, 2015). Third, much work remains to be done with respect to broadband use. As detailed in Chapter 1, the mere *availability* of broadband does not necessarily translate into *use* of these connections. Further, the range of *uses* depends on several individual level and/or business characteristics. These characteristics impact *awareness* and the *ability* to use broadband (Gibbs, 2001; Center for an Urban Future, 2004).

Conceptual framework

The question of internet use is particularly important to understanding the BEN, which is founded on the idea that technological advancement is based on the interaction of technical capability, human needs and the *unique context* in which technological advancements take place (Rosenberg, 1994). Aside from the requisite education and skills, there is also the question of how skills and education translate into productive and economically viable uses of the Internet. In this respect, several studies recognize that human capital is a critical complementary factor that plays a role in the productivity impacts of ICTs on regional economies (Entorf & Kramarz, 1998; Bartelsman & Doms, 2000; Kolko, 2010). Mack and Faggian (2013), for example, found that regional productivity increases related to broadband were only likely when both highly skilled workers and broadband were present in the same location.

The relationship between broadband and entrepreneurial activity may be summarized with the conceptual framework presented in Figure 7.1. Much like the BBN, the quadrants are relatively self-explanatory. For example, the low-low quadrant represents places with low levels of both entrepreneurial activity and broadband provision. Conversely, the high-high quadrant characterizes locales with high levels of entrepreneurial activity and broadband provision. The other two quadrants represent places with mixed profiles. If one thinks of these relationships as a sub-dimension of the BBN discussed in Chapter 6, high levels of broadband provision may positively impact entrepreneurial activity. The reverse may also be true. Of particular interest are regions with a mixed BEN. These are locations that might be overachieving or underachieving in entrepreneurial activity or broadband provision. The question is, why?

Broadband

	High	Low
High Entrepreneurial Activity	High levels of entrepreneurial activity, High levels of broadband	High levels of entrepreneurial activity, Low levels of broadband
Low	Low levels of entrepreneurial activity, High levels of broadband	Low levels of entrepreneurial activity, Low levels of broadband

Figure 7.1 The Broadband-Entrepreneurship Nexus.

An empirical evaluation of broadband and entrepreneurial activity

The remainder of this chapter provides an exploratory look at the association between broadband and entrepreneurial activity using broadband data from the Federal Communications Commission (FCC) and point-level establishment information.[4] Once again, this is not structured as a comprehensive, confirmatory, spatial analysis, nor a detailed technical treatise of the spatial clustering of entrepreneurial ventures and broadband providers. Rather, the analysis is designed to be an exploratory analysis of the association between broadband and entrepreneurial activity at a fine-grained spatial scale. While prior work has evaluated the relationship between broadband and new firm formation across states and counties using North American Industrial Classification System (NAICS) data (Parajuli & Haynes, 2012), more fine-grained analyses have yet to be conducted.

ZIP code level broadband data between 1999 and 2007 were obtained from the FCC's Form 477 database. Census tract level data were also obtained from this database from 2008 to 2013. Point-level establishment data were obtained from the National Establishment Time Series (NETS) database for select metropolitan areas between 1990 and 2012. Based on information about the start date of each establishment, it was possible to identify new establishments on an annual basis for this twenty-two-year time series. After identifying new establishments that correspond to the years for which data are also available from the FCC, these point data were aggregated to ZIP code areas and Census tracts accordingly. Mimicking previous work (Renski, 2008; Parajuli & Haynes, 2012), counts of new firms are used as the key variable in the following analysis. The reason for using counts is a simple one. The association of interest is between broadband provision and new firm presence, not growth compared to existing establishments or the size of the labor force. As a result, we do not standardize firms by the number of existing establishments, or the size of the labor force – something done with traditional ecological and labor market approaches, respectively.

Metropolitan area trends

Six metropolitan areas are analyzed in this exploratory analysis: Austin, Boston, Kansas City, Philadelphia, Raleigh–Durham and San Jose, CA. Of these metropolitan areas Austin, Boston, Raleigh and San Jose are locales with historically high levels of entrepreneurial activity. Kansas City is interesting because of the ongoing installation of Google Fiber and corresponding entrepreneurial activity that has cropped up in recent years. Philadelphia is potentially interesting because it represents a metropolitan area that has not been studied previously in the entrepreneurship literature. Philadelphia is also particularly interesting where broadband is concerned because of the city-level efforts to deploy wireless broadband (Rose, 2008).

Figures 7.2 and 7.3 present time series trends for new establishments and broadband provision in each of the eight metropolitan areas. Specifically, Figure 7.2 presents time series trends for the number of new establishments per 100,000

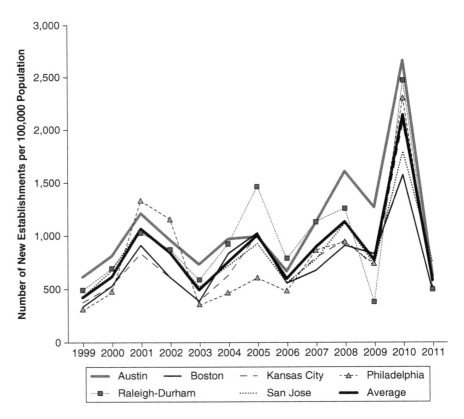

Figure 7.2 Time series trends in entrepreneurial activity, 1999–2011.

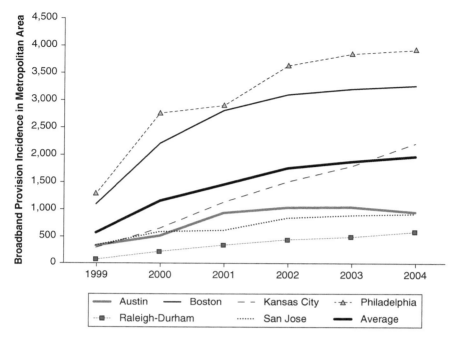

Figure 7.3 Time series trends in broadband, 1999–2004.

people between 1999 and 2011. The dark black line in this graph represents the average new establishment activity across these metropolitan areas for this same period. Of note in Figure 7.2 is that new establishment activity is increasing over time. For example, in 1999, the number of new establishments per 100,000 people was 423 compared to an average of 583 in 2011. There is a peak of entrepreneurial activity in 2010 across all six metropolitan areas. This peak is likely related to people pursuing entrepreneurship as an employment option given the negative employment impacts of the 2008 recession.

Figure 7.2 also highlights movements of each of the six metropolitan areas of interest in terms of their relative levels of new establishment activity. In 1999, the top metropolitan areas were Austin, Raleigh and San Jose. By 2011, however, the top metropolitan areas had shifted a bit. San Jose, Kansas City and Austin had the highest levels of new establishment activity for 2011. Raleigh–Durham is perhaps the most dynamic metropolitan area for new establishment activity. After virtually leading all metropolitan areas in 1999, activity in Raleigh–Durham fell in 2009 and again in 2011 to just 495 new establishments per 100,000 people. Boston is also a laggard across the entire time series, which is somewhat surprising given the amount of entrepreneurial activity in the Route 128 area (Saxenian, 1994). Prior studies have highlighted that Boston was hit particularly hard by the dot.com bust in 2001 and the 2008 recession (Woolhouse, 2008).

Table 7.2 summarizes trends in new establishment activity by splitting the six metropolitan areas into two categories, those with below average levels of entrepreneurial activity (low), and those with above average levels of entrepreneurial activity (high). It is important to note that these are relative measures and only consider the six metropolitan areas used in this analysis. In turn, this information will be fused with data concerning broadband provision levels to explore the BEN.

Figure 7.3 presents information about the total number of broadband providers in metropolitan areas between 1999 and 2004, which is a metric that has been used frequently in prior studies (Prieger, 2003; Kolko, 2010, 2012). It is important to note that these are not unique counts of providers for a region and its ZIP code areas. Rather, these numbers reflect the incidence of broadband provision, in aggregate, for all ZIP code areas in the metropolitan area. Thus, these counts should be viewed and interpreted as a relative measure of broadband provision for each region, not a unique count or inventory. The graphic is restricted to this six-year time series because of differences in reporting requirements associated with FCC Form 477 detailed in Chapter 4. That said, Boston and Philadelphia have some of the highest levels of broadband provision over the entire time series while Raleigh–Durham, San Jose and Austin are lagging. The reasons for these differences could be simple, such as the overall size of the population in each metropolitan area. Or, these differences could be a function of a more complex array of factors, such as race, income, age and local provider competition (Grubesic, 2003; Martin & Robinson, 2007; Prieger & Hu, 2008).

Table 7.3 summarizes the information about broadband provision in metropolitan areas for 1999–2011. Again, the same caveats for tabulating broadband provision incidence apply. This summary allows for a quick, comparative snapshot of provision levels, which is particularly valuable given the discontinuities associated with broadband data. As was done with the new establishment data, average levels of broadband provision are used to split metropolitan areas into two categories. Metropolitan areas with below average levels of provision are labeled as "low" and those with above average levels of provision are labeled as "high." Metropolitan areas with consistently high trends in provision are highlighted in gray (Table 7.3). While prior research has noted some dynamism to provision levels in metropolitan areas over time (Grubesic, 2008), this table reveals that provision levels in the metropolitan areas of interest are fairly static. Raleigh, and San Jose have consistently lower levels of broadband than average. Boston and Philadelphia have consistently higher levels (Figure 7.3).

Figure 7.4 merges the information from Tables 7.1 and 7.2 at four points in time to fill in the conceptual framework of the BEN for 1999, 2003, 2007 and 2011. The use of the framework reveals that this relationship is dynamic over time. It also reveals that both entrepreneurial activity and broadband provision are drivers of metropolitan areas that change positions within the framework over time. Consider, for example, Kansas City, which moved from the low/low position in 1999 to the low/high position in 2003 then back to the low/low position in 2007, before ending up in the high/low position in the final year of the study

Table 7.2 Startups per 100,000 population

	1999	2000	2001	2002	2003	2004	2005	2006	2007	2008	2009	2010	2011
Austin	617.47	815.61	1212.28	962.81	736.39	973.51	995.76	669.49	1130.46	1606.19	1269.19	2662.54	622.10
Boston	333.43	529.74	912.56	612.99	387.40	835.49	1029.05	557.99	676.87	908.76	831.44	1572.57	482.35
Kansas City	376.59	532.89	835.36	609.24	391.33	632.40	1023.69	551.73	797.71	957.93	715.64	2057.17	634.37
Philadelphia	300.92	467.48	1330.29	1158.60	350.15	464.02	607.04	480.47	867.02	942.22	731.54	2300.37	508.31
Raleigh–Durham	490.35	695.14	1032.05	872.41	587.21	927.75	1468.36	780.92	1134.10	1259.42	373.40	2474.43	495.44
San Jose	424.32	619.91	1066.74	844.67	495.81	759.25	1008.74	600.37	897.58	1132.38	775.70	2143.31	582.78
Average	**423.85**	**610.13**	**1064.88**	**843.45**	**491.38**	**765.40**	**1022.11**	**606.83**	**917.29**	**1134.48**	**782.82**	**2201.73**	**554.23**

	1999	2000	2001	2002	2003	2004	2005	2006	2007	2008	2009	2010	2011
Austin	high	high	high	high	high	high	low	high	high	high	high	high	high
Boston	low	low	low	low	low	high	high	low	low	low	high	low	low
Kansas City	low	low	low	low	low	low	high	low	low	low	low	low	high
Philadelphia	low	low	high	high	low	low	low	low	high	low	low	high	low
Raleigh–Durham	high	high	low	high	high	high	high	high	high	high	low	high	low
San Jose	high	high	high	high	high	low	high	low	low	low	low	low	high

Table 7.3 Summary of broadband provision activity

	1999	2000	2001	2002	2003	2004	2005	2006	2007	2008	2009	2010	2011
Austin	334	516	936	1037	1046	951	668	730	764	1876	2284	2279	2969
Boston	1097	2213	2816	3109	3214	3273	2473	2274	2590	5868	5807	5670	6144
Kansas City	295	653	1134	1512	1802	2215	1614	1629	1716	3689	3626	3537	3841
Philadelphia	1284	2763	2907	3642	3856	3932	3278	2941	3807	9928	10107	8733	8795
Raleigh–Durham	68	220	344	448	496	591	531	497	497	1090	1134	1072	1524
San Jose	341	590	617	845	895	918	899	663	629	2420	2447	2631	3211
Average	**569.833**	**1159.17**	**1459.00**	**1765.50**	**1884.83**	**1980.00**	**1577.167**	**1455.667**	**1667.167**	**4145.167**	**4234.167**	**3987**	**4414**

	1999	2000	2001	2002	2003	2004	2005	2006	2007	2008	2009	2010	2011
Austin	low	low	low	low	low	low	low	low	low	low	low	low	low
Boston	high	high	high	high	high	high	high	high	high	high	high	high	high
Kansas City	low	low	low	high	high	high	high	high	high	high	high	high	high
Philadelphia	high	high	high	high	high	high	high	high	high	high	high	high	high
Raleigh–Durham	low	low	low	low	low	low	low	low	low	low	low	low	low
San Jose	low	low	low	low	low	low	low	low	low	low	low	low	low

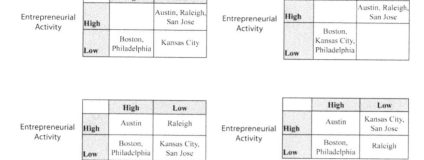

Figure 7.4 Broadband-Entrepreneurship Nexus (BEN) over time.

(2011). Boston and Philadelphia represent metropolitan areas that are fairly static in the context of this framework. They remain in the low/high position for the entire study period.

   In summary, the BEN represents a valuable tool for analyzing joint trends in broadband provision and entrepreneurial activity over time. It is important to note, however, that this framework is limited to evaluating the coincidence of trends in these two variables over time; it does not imply a causal or statistical relationship between these two variables. Therefore, it is recommended that this analytical framework be supplemented with a more rigorous confirmatory analysis. A second limitation of this exploratory analysis is the use of averages for the metropolitan areas for which NETS data are available. An advantage of this framework, however, is its flexibility. This means the framework may be used with other proxies for entrepreneurial activity that have national level coverage, such as proprietor and self-employment information from the Bureau of Economic Analysis (BEA).

Intra-metropolitan trends

Another limitation of the BEN, at least as it was detailed above, is that it is relatively aspatial. This means that the inter-metropolitan trends in broadband provision and entrepreneurial activity highlighted above, while valuable, tend to smooth important intra-metropolitan trends for the two variables of interest. Given this limitation, the use of local statistics in combination with this framework is recommended to analyze spatial trends in broadband and entrepreneurial activity. Figures 7.5 and 7.6 present intra-metropolitan patterns of broadband provision and entrepreneurial activity for Boston and Kansas at two points in time 1999 and 2011.

   The three maps for each metropolitan area displayed in Figures 7.5 and 7.6 were produced using two local statistics, the local Moran's I (Anselin, 1995) and

Figure 7.5 Spatial dimensions of the BEN for Boston and Kansas City, 1999.

the bivariate local Moran's I (Anselin, Syabri, & Kho, 2006), respectively. The motivation for analyzing the spatial patterns in broadband provision and entrepreneurial activity, separately, and then jointly via the bivariate local Moran, is to provide some sense of statistical significance between the spatial patterns for the two variables. However, as detailed with the BBN in Chapter 6, these results must be interpreted with caution. The bivariate approach omits the central observation of interest in the calculation of the statistic. Thus, while it is possible to get some sense of the statistical significance of the joint spatial patterns between two variables it is not necessarily possible to say that entrepreneurial activity clusters around spatial units with high levels of broadband provision, or vice versa. Limitations aside, all local tests were computed with the exact same spatial weights to facilitate comparisons for each year.[5]

Figure 7.5 highlights that in 1999, broadband provision in Boston was clustered in the central city, and radiated in a primarily northwesterly direction toward Route 128 and its suburbs, including Concord, Acton and as far north as Lowell. New establishment activity was also clustered in the same general region, including the cities of Cambridge, Somerville and Arlington. Not surprisingly, the results for the bivariate measure of spatial autocorrelation largely mimic that of the univariate measures, suggesting that hotspots of high broadband and new establishment activity are focused on central Boston and its northwestern suburbs. For this particular metropolitan area and year then, it appears that broadband provision is associated with new establishment activity.

Figure 7.6 Spatial dimensions of the BEN for Boston and Kansas City, 2011.

The results for Kansas City, however, present a much different picture (Figure 7.5). For 1999, there are no strongly identifiable hotspots of broadband activity; rather, there are pockets of ZIP code areas in suburban and exurban locales with high levels of provision, relative to their neighbors. This is perhaps related to the income level of households in these locations and their ability to pay for broadband when it was still a relatively novel way to connect to the Internet (Grubesic, 2008). New establishment activity, however, is primarily concentrated in the suburbs south of Kansas City, KS (not its larger neighbor in Missouri), and includes cities such as Westwood, Roeland Park and Prairie Village. The bivariate analysis displays a somewhat different result and reveals that higher broadband provision levels are not associated with new establishment activity. This suggests that during 1999, broadband remained a have/have-not issue for the region, rather than one of provider quantity and/or competition. It may also be the case that the types of new establishments emerging in Kansas City are in different industries than those emerging in Boston, which is a well-known hub for innovation in electronics (Saxenian, 1994).

Just over a decade later, Figure 7.6 displays very different spatial patterns in both metropolitan areas. In Boston, for example, broadband provision exhibits a core-periphery pattern, with a strong presence in Boston and its inner-ring suburbs. There is also a hotspot of broadband provision in/around Portsmouth, NH. A similar core-periphery pattern is revealed in the bivariate analysis. However, Boston's Route 128 corridor, long a hub of electronics related innovation

(Rosegrant & Lampe, 1992; Saxenian, 1994), is identified as an area where high levels of broadband provision are associated with high levels of new establishment activity. Where Kansas City is concerned, Figure 7.6 displays a markedly different spatial pattern. Unlike Boston, there is no core-periphery pattern. Instead, hotspots of broadband provision and new establishments appear to be clustered in and around the suburbs of Lenexa and Overland Park. Not surprisingly, the Sprint Corporation, one of the world's largest telecommunication providers, has its global headquarters in Overland Park, employing close to 8,000 people (Sprint, 2015). Netsmart Technologies, a large health information technology company, is also located in the region. Given the synergies created by these large IT companies, dozens of other large corporations, their many employees and the presence of the University of Kansas (located in Lawrence, KS), it is not surprising that high levels of broadband and new establishments cluster in this region.

Conclusion

The analysis highlighted above is an exploratory mechanism for analyzing the spatial relationships between broadband provision and entrepreneurial activity in the form of new establishments. The status of broadband internet connections as a GPT has produced numerous downstream complementarities in the form of innovations in devices and applications. These innovations have yielded some of the most revolutionary and popular devices of our time, including the smartphone and tablet computers. In addition, application innovations, in the form of email, social media, application marketplaces and development platforms such as Java have permanently altered the way in which humans interact and communicate. Combined, these innovations suggest a link between broadband availability and entrepreneurial activity. Surprisingly, while scholars agree that technological change is one of the driving forces behind economic growth, broadband, which is one of the more recent instances of technological change, remains an understudied aspect of the entrepreneurship literature.

Given the association between broadband and entrepreneurial activity, this chapter designed a conceptual framework, which represents a sub-dimension of the BBN, entitled the Broadband-Entrepreneurship Nexus. Based on this framework, an exploratory analysis of the association between broadband provision and entrepreneurial activity was conducted for six metropolitan areas in the United States. The results suggested that broadband provision was not always tethered to entrepreneurial activity within metropolitan areas. However, there were several locales, including Boston and ultimately Kansas City, where this association was important. In addition, the linkage between broadband and entrepreneurial activity appears to be dynamic in both space and time. While more work is needed to statistically evaluate the intra-metropolitan association between broadband and entrepreneurial activity, this study highlights the vibrant and highly contextual nature of this relationship. Usage evaluations of broadband by entrepreneurs and the larger business community are also necessary to understand the nuances in the BEN. This need and several others are discussed in the closing chapter of the book.

Notes

1 See Greenstein and Prince (2006) for citations.
2 The first mobile phone call was made on a Motorola DynaTAC 8000x on April 3, 1973 (Seward, 2013). The DynaTAC 8000x is also considered to be the world's first cellphone (ibid.).
3 For readers who are software geeks, it is important to note that Android uses Java, but does not use its class libraries. Java code is executed with the Dalvik virtual process machine (VM), instead of the Java VM. More importantly, the Dalvik VM is rapidly being replaced by Android Runtime (Anthony, 2013)
4 Establishments are defined as a single physical location at which business takes place (U.S. Census Bureau, 2015).
5 These figures were produced using a K-nearest neighbor spatial weights matrix with a value of 4. Sensitivity analysis was performed with alternative weight matrices, but $k = 4$ provided the most conservative and meaningful results.

References

Aalto, L., Göthlin, N., Korhonen, J., & Ojala, T. (2004). Bluetooth and WAP push based location-aware mobile advertising system. In *Proceedings of the 2nd international conference on mobile systems, applications, and services* (pp. 49–58). ACM.

Abbate, J. (1999). *Inventing the Internet*. Cambridge: The MIT Press.

Acs, Z.J., Audretsch, D.B., Braunerhjelm, P., & Carlsson, B. (2004). The knowledge filter and entrepreneurship in endogenous growth. Discussion Papers on Entrepreneurship, *Growth and Public Policy*, Paper #0805.

Acs, Z.J., & Armington, C. (2004). The impact of geographic differences in human capital on service firm formation rates. *Journal of Urban Economics*, *56*(2), 244–278.

Aghion, P., & Howitt, P. (1992). A model of growth through creative destruction. *Econometrica*, *60*(2), 323–351.

Anselin, L. (1995). Local indicators of spatial association-LISA. *Geographical Analysis*, *27*(2), 93–115.

Anselin, L., Syabri, I., & Kho, Y. (2006). GeoDa: An introduction to spatial data analysis. *Geographical Analysis*, *38*(1), 5–22.

Anthony, S. (2013). Android ART: Google finally moves to replace Dalvik, to boost performance and battery life. *ExtremeTech*, Retrieved from http://tinyurl.com/pa7qfls

Armington, C., & Acs, Z.J. (2002). The determinants of regional variation in new firm formation. *Regional Studies,* *36*(1), 33–45.

Arrow, K.J. (1962). The economic implications of learning by doing. *The Review of Regional Studies*, *29*(3), 155–173.

Ashton, K. (2009). That 'internet of things' thing. *RFiD Journal*, *22*(7), 97–114.

Audretsch, D.B., & Fritsch, M. (1994). The geography of firm births in Germany. *Regional Studies*, *28*(4), 359–365.

Audretsch, D., & Fritsch, M. (2002). Growth regimes over time and space. *Regional Studies,* *36*, 113–124.

Audretsch, D. B., & Keilbach, M. (2008). Resolving the knowledge paradox: Knowledge-spillover entrepreneurship and economic growth. *Research Policy*, *37*(10), 1697–1705.

Bailey, M., & Gordon, R. (1988). The productivity slowdown, measurement issues and the explosion of computer power. *Brookings Papers on Economic Activity*, *19*, 347–420.

Baldwin, J.R., Diverty, B. & Sabourin, D. (1995). Technology use and industrial transformation: Empirical perspective. Working Paper No. 75, Microeconomics Analysis Division, Statistics Canada, Ottawa.

Barro, R.J., & Sala-i-Martin, X. (1992). Convergence. *Journal of Political Economy*, *100*(2), 223–251.

Bartel, A.P., & Lichtenburg, F.R. (1987) The comparative advantage of educated workers in implementing new technologies. *Review of Economics and Statistics*, *69*(1), 1–11.

Bartelsman, E.J., & Doms, M. (2000). Understanding productivity: Lessons from longitudinal microdata. *Journal of Economic Literature*, *38*(3), 569–594.

Benhabib, J., & Spiegel, M.M. (1994). The role of human capital in economic development: Evidence from aggregate cross-country data. *The Journal of Monetary Economics*, *34*,143–173.

Berners-Lee, T., Fischetti, M., & Dertouzos, M.L. (2000). *Weaving the Web: The original design and ultimate destiny of the World Wide Web by its inventor*. New York: Harper Information.

Bever, L. (2014). 'Text neck' is becoming an 'epidemic' and could wreck your spine. *Washington Post*. Retrieved from http://tinyurl.com/mjmx6lw

Bresnahan, T. (2010). General purpose technologies. *Handbook of the Economics of Innovation*, *2*, 761–791.

Bresnahan, T.F., & Trajtenberg, M. (1995). General-purpose technologies 'Engines of growth'? *Journal of Econometrics*, *65*(1), 83–108.

Businessweek. (1996). *Just over the horizon space age cellulars*. July 21, 1996. Retrieved from http://tinyurl.com/on6jrg3

Campi, M.T.C., Blasco, A.S., & Marsal, E.V. (2004). The location of new firms and life cycle of industries. *Small Business Economics*, *22*(3/4), 265–281.

Center for an Urban Future. (2004). *New York's broadband gap*. Retrieved from https://nycfuture.org/research/publications/new-yorks-broadband-gap

David, P.A., & Wright, G. (1999). General purpose technologies and surges in productivity: Historical reflections on the future of the ICT revolution. *Discussion Papers in Economic and Social History*, *31*, September 1999.

Delgado, M., Porter, M.E., & Stern, S. (2010). Clusters and entrepreneurship. *Journal of Economic Geography*, *10*(4), 495–518.

Emarketer. (2014). *2 billion consumers worldwide to get smart(phones) by 2016*. December 11, 2014. Retrieved from http://www.emarketer.com/Article/2-Billion-Consumers-Worldwide-Smartphones-by-2016/1011694

Entorf, H., & Kramarz, F. (1998). The impact of new technologies on wages: Lessons from matching panels on employees and on their firms. *Economics of Innovation and New Technology*, *5*(2–4), 165–198.

Federal Communications Commission [FCC]. (2010). *Connecting America: The National Broadband Plan*. Retrieved from http://download.broadband.gov/plan/national-broadband-plan.pdf

Felsenstein, D., & Fleischer, A. (2002). Small–scale entrepreneurship and access to capital in peripheral locations: An empirical analysis. *Growth and Change*, *33*(2), 196–215.

Forbes. (2014). The world's most valuable brands. Retrieved from http://www.forbes.com/powerful-brands/list/

Gambardella, A., & McGahan, A.M. (2010). Business-model innovation: General purpose technologies and their implications for industry structure. *Long range planning*, *43*(2), 262–271.

Gibbs, D. (2001). Harnessing the information society? European Union policy and information and communication technologies. *European Urban and Regional Studies*, 8(1), 73–84.

Glaeser, E., & Kerr, W.R. (2009). Local industrial conditions and entrepreneurship: How much of spatial distribution can we explain? *Journal of Economics & Management Strategy*, 18(3), 623–663.

Goodrich, R. (2013). *Location-based services: Definition & examples*. October 30, 2013. Retrieved from http://www.businessnewsdaily.com/5386-location-based-services.html

Graham, S. (1999). Global grids of glass: On global cities, telecommunications and planetary urban Networks. *Urban Studies*, 36, 929–949.

Graham, S. (2002). Bridging urban digital divides? Urban polarisation and information and communications technologies (ICTS). *Urban Studies*, 39(1), 33–56.

Greenstein, S., & Prince, J. (2006). *The diffusion of the Internet and the geography of the digital divide in the United States* (No. w12182). National Bureau of Economic Research.

Gretton, P., Gali, J., & Parham, D. (2004). *The effects of ICTs and complementary innovations on Australian productivity growth*. Paper presented to the Workshop on ICT and Business Performance, OECD, Paris on 9 December 2002. Retrieved from http://www.pc.gov.au/research/confproc/uiict/index.html.

Griliches, Z. (1979). Issues in assessing the contribution of R&D to productivity growth. *Bell Journal of Economics*, 10(1), 92–116.

Grubesic, T.H. (2003). Inequities in the broadband revolution. *The Annals of Regional Science*, 37(2), 263–289.

Grubesic, T.H. (2008). The spatial distribution of broadband providers in the United States: 1999–2004. *Telecommunications Policy*, 32(3), 212–233.

Grubesic, T.H. (2015). The broadband provision tensor. *Growth and Change*, 46(1), 58–80.

Grubesic, T.H., &Murray, A.T. (2002). Constructing the digital divide: Spatial disparities in broadband access. *Papers in Regional Science*, 81(2), 197–221.

Grubesic, T.H., & Murray, A.T. (2004). Waiting for broadband: Local competition and the spatial distribution of advanced telecommunication services in the United States. *Growth and Change*, 35(2), 139–165.

Grush, A. (2012). *IBM Simon: World's first smartphone is now 20 years old*. November 26, 2012. Retrieved from http://www.androidauthority.com/ibm-simon-birthday-134255/

Guesnier, B. (1994). Regional variations in new firm formation in France. *Regional Studies*, 4(4), 347–358.

Helpman, E., & Rangel, A. (1998). *Adjusting to a new technology: Experience and training*. National Bureau of Economic Research (NBER) Working Paper 6551. Retrieved from http://www.nber.org/papers/w6551

Helpman, E., & Trajtenberg, M. (1994). *A time to sow and a time to reap: Growth based on general purpose technologies* (No. w4854). National Bureau of Economic Research.

Helpman, E., & Trajtenberg, M. (1996). *Diffusion of general purpose technologies* (No. w5773). National Bureau of Economic Research.

Hess, K. (2013). Three billion devices run Java. Yeah, but do they like it?. *ZDNet*. Retrieved from http://tinyurl.com/mucowjb

Jimeniz, E., & Greenstein, S. (1998). The emerging internet retailing market as a nested diffusion process. *International Journal of Innovation Management*, 2(03), 281–308.

Jovanovic, B., & Rousseau, P.L. (2002). Moore's law and learning by doing. *Review of Economic Dynamics*, 5(2), 346–375.

Jovanovic, B., & Rousseau, P.L. (2005). General purpose technologies. *Handbook of Economic Growth*, *1*, 1181–1224.

Kelly, H. (2014). Helping 'smart' devices talk to each other. *CNN*. March 28, 2014. Retrieved from http://tinyurl.com/o8lg2ql

Kolko, J. (2010). A new measure of U.S. residential broadband availability. *Telecommunications Policy*, *34*(3), 132–143.

Kolko, J. (2012). Broadband and local growth. *Journal of Urban Economics*, *71*(1), 100–113.

Lamm, S.E., Reed, D.A., & Scullin, W.H. (1996). Real-time geographic visualization of World Wide Web traffic. *Computer Networks and ISDN Systems*, *28*(7), 1457–1468.

Laugesen, J., & Yuan, Y. (2010, June). What factors contributed to the success of Apple's iPhone? In *Proceedings of the 2010 Ninth International Conference on Mobile Business and 2010 Ninth Global Mobility Roundtable* (ICMB-GMR) (pp. 91–99). Washington DC: IEEE Computer Society.

Leigh, N.G., & Blakely, E.J. (2013). *Planning local economic development: Theory and practice* (5th ed.). Thousand Oaks, CA: Sage Publications Incorporated.

Lindgardt, Z., Reeves, M., Stalk, G., & Deimler, M.S. (2009). *Business model innovation*. The Boston Consulting Group. Retrieved from http://tinyurl.com/q58y59o

Lipsey, R.G., Carlaw, K.I., & Bekar, C.T. (2005). *Economic transformations: General purpose technologies and long-term economic growth*. Oxford University Press.

Lloyd, R., & Clark, T.R. (2001). The city as an entertainment machine. In K.F. Gotham (ed.), *Critical perspectives on urban redevelopment* (Research in Urban Sociology, Volume 6; pp. 357–378). Bingley, UK: Emerald Group Publishing Limited.

Lucas, R.E. (1988). On the mechanics of economic development. *Journal of Monetary Economics*, *22*, 3–42.

Mack, E.A., & Faggian, A. (2013). Productivity and broadband: The human factor. *International Regional Science Review, 36*(3), 392–423.

Malecki, E.J. (1994). Entrepreneurship in regional and local development. *International Regional Science Review*, *16*(1–2), 119–153.

Mankiw, N.G., Romer, D., & Weil, D.N. (1992). A contribution to the empirics of economic growth. *The Quarterly Journal of Economics*, *107*(2), 407–437.

Markoff, J. (2005). *What the dormouse said: How the sixties counterculture shaped the personal computer industry*. New York: Penguin.

Martin, S.P., & Robinson, J.P. (2007). The income digital divide: Trends and predictions for levels of Internet use. *Social Problems*, *54*(1), 1–22.

Martin, T. (2014). *The evolution of the smartphone*. Retrieved from http://pocketnow.com/2014/07/28/the-evolution-of-the-smartphone

McCarty, B. (2011). The history of the smartphone. *The Next Web*. December 6, 2011. Retrieved from http://thenextweb.com/mobile/2011/12/06/the-history-of-the-smartphone/

Morgan, J. (2014). A simple explanation of 'The Internet of things'. *Forbes*. May 13, 2014. Retrieved from http://www.forbes.com/sites/jacobmorgan/2014/05/13/simple-explanation-internet-things-that-anyone-can-understand/,

Morrison, C.J. (1997). Assessing the productivity of information technology equipment in U.S. manufacturing industries. *The Review of Economics and Statistics*, *79*(3), 471–481.

Moss, M.L., & Townsend, A.M. (2000). The Internet backbone and the American Metropolis. *The Information Society*, *16*(1), 35–47.

Moschovitis, C.J., Poole, H., & Senft, T.M. (1999). *History of the Internet: A chronology, 1843 to the present*. Santa Barbara, CA: AB C-CLIO, Incorporated.

Murphy, K.M., Riddell, W.C., & Romer, P.M. (1998). *Wages, skills, and technology in the United States and Canada* (No. w6638). National Bureau of Economic Research.

National Telecommunications and Information Administration [NTIA]. (2014). *Broadband technology opportunities program (BTOP). Quarterly program status report*. Retrieved from http://tinyurl.com/kokdquh

O'Kelly, M.E., & Grubesic, T.H. (2002). Backbone topology, access, and the commercial Internet. *Environment and Planning B, 29*(4), 533–552.

Oracle. (2015). *Design goals of the Java programming language*. Retrieved from http://tinyurl.com/mfwz2mh

Parajuli, J., & Haynes, K.E. (2012). *Broadband Internet and new firm formation: A U.S. perspective*. George Mason School of Public Policy Research Paper No. 2013-03.

Perera, C., Zaslavsky, A., Christen, P., & Georgakopoulos, D. (2014). Sensing as a service model for smart cities supported by internet of things. *Transactions on Emerging Telecommunications Technologies, 25*(1), 81–93.

Perez, S. (2014). iTunes App Store now has 1.2 million apps, has seen 75 billion downloads to date. *TechCrunch*. Retrieved from http://tinyurl.com/kowlhq6

Pew Research Center. (2014). *Mobile technology fact sheet*. Retrieved from http://www.pewinternet.org/fact-sheets/mobile-technology-fact-sheet/

Press, G. (2014). Internet of things by the numbers: Market estimates and forecasts. *Forbes*. August 22, 2014. Retrieved from http://tinyurl.com/l38j2yk

Prieger, J.E. (2003). The supply side of the digital divide: Is there equal availability in the broadband Internet access market? *Economic Inquiry, 41*(2), 346–363.

Prieger, J.E., & Hu, W.M. (2008). The broadband digital divide and the nexus of race, competition, and quality. *Information Economics and Policy, 20*(2), 150–167.

Renski, H. (2008). New firm entry, survival, and growth in the United States: A comparison of urban, suburban, and rural areas. *Journal of the American Planning Association, 75*(1), 60–77.

Roach, S. (1987). *America's technology dilemma: A profile of the information economy*. New York, NY: Morgan Stanley.

Rogers, Y. (2009). *The changing face of human-computer interaction in the age of ubiquitous computing* (pp. 1–19). Berlin Heidelberg: Springer.

Romer, P.M. (1986). Increasing returns and long-run growth. *Journal of Political Economy, 94*(5), 1002–1037.

Romer, P.M. (1990). Endogenous technological change. *Journal of Political Economy, 98*(5), S71–S102.

Rose, J. (2008). Philly fears Earthlink may bail on WiFi network. National Public Radio. Retrieved from http://tinyurl.com/lw6oevm

Rosegrant, S., & Lampe, D.R. (1992). *Route 128: Lessons from Boston's high-tech community*. New York: Basic Books, Inc.

Rosenberg, N. (1994). *Exploring the black box: Technology, economics, and history*. Cambridge University Press.

Saxenian, A. (1994). *Regional advantage: Culture and competition in Silicon Valley and Route 128*. Cambridge, MA: Harvard University Press.

Schiller, J., & Voisard, A. (Eds.). (2004). *Location-based services*. Elsevier.

Seward, Z.M. (2013). The first mobile phone call was made 40 years ago today. *The Atlantic*. Retrieved from http://tinyurl.com/c3bduhy

Shaughnessy, H. (2013). Who has the winning innovation model, Google, Apple, or Samsung? *Forbes*. 3/7/2013. Retrieved from http://tinyurl.com/buyo69m

Solow, R.M. (1956). A contribution to the theory of economic growth. *The Quarterly Journal of Economics, 70*(1), 65–94.

Sprint. (2015). *The story of Sprint.* Retrieved from http://www.sprint.com/companyinfo/history/

Statista. (2014). Number of apps available in leading app stores as of July 2014. Retrieved from http://tinyurl.com/nrr5sxo

Storey, D. (1991). The birth of new firms? Does unemployment matter? A review of the evidence. *Small Business Economics, 3*(3), 167–178.

Strausman, P.A. (1990). *The business value of computers.* New Canaan, CT: Information Economics Press.

Strover, S. (2001). Rural internet connectivity. *Telecommunications Policy, 25*, 331–347.

Tödtling, F., & Wanzenböck. H. (2003). Regional differences in structural characteristics of start-ups. *Entrepreneurship & Regional Development, 15*(4), 351–370.

U.S. Census Bureau. (2015). Statistics of U.S. businesses. Definitions. Retrieved from https://www.census.gov/econ/susb/definitions.html

Wagener, J., & Sternberg, R. (2004). Start-up activities, individual characteristics, and the regional milieu: Lessons for entrepreneurship support policies from German micro data. *Annals of Regional Science, 38*, 219–240.

Wennekers, A.R.M., & Thurik, A.R. (1999). Linking entrepreneurship and economic growth. *Small Business Economics, 13*(1), 27–55.

Woolhouse, M. (2008). Mass. has regained jobs lost in last recession. *Boston Globe.* March 22, 2013. Retrieved from http://tinyurl.com/qj2ofhf

Xing, Y., & Detert, N.C. (2010). How the iPhone widens the United States trade deficit with the People's Republic of China. Retrieved from http://papers.ssrn.com/sol3/papers.cfm?abstract_id=1729085

Young, A. (1993a). Invention and bounded learning by doing. *Journal of Political Economy, 101*(3), 443–472.

Young, A. (1993b). Substitution and complementarity in endogenous innovation. *The Quarterly Journal of Economics, 108*(3), 775–807.

Zook, M.A. (2003). Underground globalization: Mapping the space of flows of the Internet adult industry. *Environment and Planning A, 35*(7), 1261–1286.

8   The future of broadband

Predicting the future is a risky business. This is especially true when it comes to technology. Sieber (2011) outlines a number of inaccurate predictions about computer technology over the years. For example, in 1977, Ken Olsen, the founder of Digital Equipment Corporation, stated, "There is no reason anyone would want a computer in their home." Twelve years later in 1989 Bill Gates (Microsoft) stated, "We will never make a 32-bit operating system." Windows NT, a 32-bit operating system, was released by Microsoft in 1993.

As authors of this book, we are not in the business of making technology predictions and this is especially true when it comes to broadband. However, this will not prevent a prospective look forward at several key challenges that remain for broadband and its ability to advance regional competitiveness. Challenges that we discuss include: 1) infrastructure deployment, 2) data reporting and availability, 3) broadband use, 4) broadband policy and 5) broadband in the developing world.

Infrastructure deployment

Chapter 3 of this book, which deals with infrastructure, highlights the variety of technologies and platforms that enable broadband, as well as the problems and prospects with deploying and upgrading broadband infrastructure. Chapter 5 focuses on the geographic distribution of this infrastructure. What is clear from this analysis and related work by Grubesic (2015), is that broadband is widely deployed in the United States. However, there are significant variations in the quality of available broadband service, particularly with regards to speed. Many users in the U.S. suffer with gaps between realized and advertised speeds (Grubesic, 2015; NBM, 2015). Although recent surveys suggest that customers are relatively satisfied with their broadband service (JD Power, 2013), consumer ignorance should not be a license for providers to overcharge and under-deliver.

Although many communities throughout the United States have access to some form of broadband service, the quality of these services varies. This is particularly true for exurban, rural and remote locations that continue to grapple with broadband deployment issues. These patterns in service quality variation have persisted over time. Large, densely populated urban areas (e.g., New York, Los Angeles and Seattle) are the first locations to receive significant network upgrades and the rollout of next-generation technologies. Exurban and rural areas often find themselves at least one generation behind urban locales (if they are lucky), and the most geographically remote communities in the United States are often two generations or more behind their urban counterparts. These persistent gaps can be attributed to the propensity of private infrastructure providers to seek faster returns on investment, thereby catering to markets with higher densities of consumers and/potential subscribers (Grubesic & Murray, 2002, 2004; O'Kelly & Grubesic, 2002). There is no reason to expect that these patterns will change any time soon.

As bandwidth consumption continues to grow, particularly as the developed world moves toward an internet of things, service quality will become more important. As a result, the bulk of future deployment efforts will likely focus on systemic upgrades that provide more symmetrical bandwidth (i.e., upload/download) and the minimization of downtimes and disruptions. These needs will preclude some of the existing, widely deployed and adopted technologies for terrestrial broadband. For example, older technologies such as ADSL will be phased out in favor of newer technologies such as VDSL2 and optical fiber. It also means that the operational characteristics of fiber will make it the dominant broadband platform of the future. Over time, the costs to produce fiber will decline, the technologies for deploying broadband over fiber will improve and the methods to install fiber will become less invasive and more cost effective. Consider the process required for installing fiber in 2015. Streets are cut, earth is trenched and excavated, fiber is threaded through the trenches, earth is replaced and streets are patched and repaved. In the future, the installation of fiber will become more surgical and minimally invasive. Existing infrastructure systems will be used, including pipe networks in urban areas (e.g., water and sewer) and fiber will be installed by robotic drones. In locations where this type of existing network infrastructure is not available, minimally invasive plowing or aerial approaches will remain cost effective for fiber installations.

So, where does this leave wireless? There are proponents of wireless technology that suggest the death of wired broadband is beginning (Worstall, 2013). This viewpoint is misguided. Wireless and wireline broadband are symbiotic and complementary; they will remain this way for many years to come. There are several reasons for this. First, all cellular towers, Wi-Fi hubs and other wireless systems are reliant on wireline systems for data transit and routing. In short, wireless cannot currently exist without wireline. Second, licensed spectrum is in short supply (VantagePoint, 2015), leaving providers with few options for accommodating a high density of customers via wireless channels. Third, wireless systems are susceptible to interference from complex topography (e.g., mountains and hills), vegetation (e.g., trees) and weather (e.g., rain, snow and fog). In contrast, wireline systems are not affected by these environmental conditions. Finally, there is simply not enough bandwidth in wireless systems to support existing consumer demand (VantagePoint, 2015).

In sum, the dream of ubiquitous broadband will remain a dream until the pervasive cost issues associated with wireline technologies can be addressed. We are confident that this will occur, at some point. However, the spotty geographic coverage associated with wireless technologies, and their associated problems with quality of service (QoS) and data speeds, are likely to remain a problem for some time, particularly in rural and remote areas. As a result, while it is technically possible to achieve ubiquitous internet availability, there is a variety of technological, geographic, social, cultural and economic barriers to realizing this reality (Wilson & Corey, 2011). Instead, the technologies that provide people and businesses with internet access are likely to remain a mosaic, with significant variations across the urban–rural hierarchy.

Broadband data

A key step in analyzing the gaps in coverage associated with this mix of technologies will rely on more accurate and exhaustive data collection efforts. Chapter 4 highlights the problems and prospects with currently available data sources, particularly the recently composed National Broadband Map (NBM), a joint effort by the Federal Communications Commission (FCC) and the National Telecommunications and Information Administration (NTIA). Despite government efforts to improve broadband data availability, the NBM is fraught with problems that include: varying provider participation in data collection initiatives at the state level, differences in data collection and tabulation procedures between states, inaccuracies in broadband coverage information, and tabulation issues related to the switch from 2000 Census geographies to 2010 Census geographies. This change in geography proved particularly problematic for the June 2011 version of the NBM. The alternative source of data for broadband information, the FCC Form 477 database, proved to be somewhat more stable for analytical purposes, but these data are not without their own issues. For several years, these data were collected using ZIP code boundaries as the base geography, which are problematic for spatial analysis (Grubesic & Matisziw, 2006; Grubesic, 2008). Reporting requirements and base geographies are also dynamic over time. This renders these data unusable in long panel studies of broadband availability.

Unfortunately then, despite efforts to provide the research and policy community with good broadband data, government efforts have largely failed. The data are inaccurate, imprecise and poorly structured. We fully understand that this will be an unpopular statement, especially with all the scientists and analysts who have worked tirelessly to bring data repositories such as the NBM to life. However, this statement is true and the results detailed throughout this book empirically support it.

This begs the question, why do publicly available broadband datasets remain of such low quality? To provide some perspective on this issue, consider the events of June 2006, when researchers from the Pew Internet & American Life project, University of Texas-Austin and the Massachusetts Institute of Technology convened a workshop in Washington, DC to discuss challenges involving the collection of broadband data and the deployment and use of communications infrastructure (Flamm, Friedlander, Horrigan, & Lehr, 2006). Three major recommendations were made by the participants of the meeting:

1   The collection of broadband data should be at a sufficiently fine-grained level to permit regional analysis of the impacts of telecommunication technology.
2   The U.S. should be able to produce a map showing the availability of infrastructure in the country.
3   Academic researchers, non-profit organizations, the government and the private sector must work collaboratively to gather data that permits assessment of QoS and the user experience.

Nearly ten years later, all three of these goals have been achieved, at least to some extent. Broadband data are collected at the block level. This is fine-grained. There is an NBM for the U.S. Researchers and non-profits do interact with the government and its efforts to measure and monitor broadband, as does the private sector. Again, the NBM is a good example of this collaboration. Unfortunately, as Ashton (2009, 1) notes, "people have limited time, attention and accuracy—all of which means they are not very good at capturing data about things in the real world." In this spirit, we believe there are three areas in particular that need to be improved upon in future broadband data collection efforts.

First it is vital to collect and report pricing information. Nearly ten years after the meeting in Washington, DC, the broadband community knows nothing about price. Providers will not divulge this information and the government will not force them to disclose market pricing. However, there are ways to report this information without compromising the pricing strategies of providers. For example, information collected at the block group level (which is a fine-grained level of geography but not so fine-grained that the identity of providers is revealed) could report the average price of broadband connections as well as the minimum and maximum price for local services. In the reporting of information, care should be taken to distinguish between prices for varying speeds of broadband as well how broadband prices are embedded within triple play packages that combine telephone, cable and internet service together (Green, 2002; Schilke & Wirtz, 2012).

A second area of broadband data collection that can be improved is the reporting of speed information. As detailed previously, "advertised" speeds are very rarely an accurate reflection of realized speeds. Chapter 3 highlights a variety of reasons why this may occur from an infrastructure standpoint. One example given was the issue of power users of hybrid fiber-coaxial cable (HFC) lines. When available bandwidth is pooled for a local area, power users that consume more than their share of bandwidth can slow down the performance of other users assigned to that node. This is more likely at some times of day than others (Matisziw, Grubesic, & Guo, 2012) and is similar to peak-usage times of electricity. Advances in automated speed tests from sources such Ookla and the FCC white box program have helped provide more resolution deviations in realized speeds from advertised speed, but participation rates remain uneven. Thus, strategies for obtaining a more complete census of broadband performance in terms of speed are required and are now possible.

A third area where broadband data can be improved is error and uncertainty mitigation. Despite the NTIA's best efforts, the data generated by the state broadband program and its associated agencies remain full of problems. The source of many of these problems is divergent measurement and data collection efforts. This was and continues to be one of the primary issues with the NBM. Each state had variations in provider participation as well as the data collection process used by designated entities. In particular, the strategies used to transition from Census 2000 to Census 2010 geographies varied considerably. We believe that many of these disconnects stem from the decentralized manner in which broadband data are/were collected. Perhaps this was necessary to create a national map, but future data collection efforts for the NBM should have a more extensive level of federal oversight and coordination of state level entities. This oversight team should have extensive training in geography, spatial analysis and geocomputation. As things stand now, it is also imperative that funds and effort be allocated to fixing the historical information that is available in the NBM. Again, the team that undertakes this task should have deep experience in data wrangling, geocomputation and the spatial sciences.

One alternative strategy for collecting information about speed and price, may involve citizen participation in data collection. Participatory efforts in mapping city data and data collection are becoming increasingly popular (Hartman, 2015). Detroit for example, is making increasing use of citizen mapping and data collection efforts in their GODATA (Government Open Data Access To All) database and their Detroit Food Mapping (DFM) initiatives (Hartman, 2015). Volunteered geographic information has also proved valuable in disaster relief efforts in Haiti (Zook, Graham, Shelton, & Gorman, 2010). Chapter 4 highlights that consumer reporting of coverage issues and dead zones via outlets such as deadcellzones.com and OpenSignal.com may serve as promising complementary tools for ground truthing the broadband data reported by the FCC and NTIA.

Broadband use

Broadband use continues to be a blind spot for telecommunications policy. In this book, use is addressed indirectly. Chapter 6 discusses the Broadband-Business Nexus (BBN), highlighting the importance of considering the impact of broadband on business location, and the impact that businesses may play in fostering demand for broadband. Similarly, Chapter 7 discusses a sub-dimension of the BBN, the Broadband-Entrepreneurship Nexus (BEN), which is related to the new business opportunities that complementary innovations in broadband related devices and applications have created. Combined, these chapters highlight the important role that broadband plays in business competitiveness. While these two topics merit additional research, we strongly suggest that a focal point of this future research be an evaluation of questions regarding *broadband use*.

Unraveling business use of broadband is a complex task because it requires consideration of a confluence of factors including firm size, industry membership and employee composition. Although little research exists about broadband use and its contribution to business competitiveness, some work provides important clues to it potential impacts. Prior work has found that the use of social media helps entrepreneurs market products and services, and receive feedback from customers (Van Der Krogt, 2011). A case study of South African entrepreneurs finds that social media use is an important opportunity for building online social capital (Stevens, 2013).

In addition to firm-specific characteristics such as size and industry membership, the education and skill level of firm employees are also likely to play a role in any productivity impacts realized by business use of broadband. This hypothesis is based on the results of technology adoption studies, which find a positive link between human capital, technology adoption (Benhabib & Spiegel, 1994) and technology-related productivity impacts (Baldwin, Diverty, & Sabourin 1995; Entorf & Kramarz 1998; Gretton, Gali, & Parham 2004). With respect to broadband specifically, recent work has found evidence of skills-biased technological change in regional productivity levels (Mack & Faggian, 2013).

In-depth evaluations of broadband use are important for unraveling when, where, how and why technology contributes to firm productivity. While it is largely presumed that broadband enhances firm productivity, there are reasons to suspect the reverse may be true. The availability of applications and internet content may prove to be distractions in the workplace if employees use the Internet for activities besides work. Evaluations of broadband application use, for example, find that a switch from dial-up to broadband internet connections increased very specific types of application use such as online shopping and music acquisition rather than more productive uses such as job, career and government website use (Kolko, 2010).

To date, empirical work in this domain presents conflicting evidence about internet use and worker productivity. A study by Websense (2005) found that misuse of the Internet at work costs U.S companies $178 billion dollars annually, which translates into $5,000 per worker per year. Alternatively, studies have found that activities such as Workplace Internet Leisure Browsing (WILB) increase productivity because they allow workers a small mental break, which temporarily takes their mind off work (Cheng, 2009). A more recent study by Pew Research found that 46% of workers felt the Internet made them more productive; citing that it expanded the number of people they worked with, increased the number of hours worked and allowed them greater flexibility (Purcell & Rainie, 2014). This study cited differences in the importance of online based tools between office and non-office workers. Office workers[1] felt tools such as email, the Internet and landline phone were more important than did non-office workers. Non-office workers felt smartphones were relatively more important than office workers (Purcell & Rainie, 2014). A recommendation for business surveys is that they include information about industry membership, as well as employee education and skill levels. This type of information can not only deepen our understanding of business specific productivity impacts, it could also reveal the type of training that workers need to make better use of internet applications.

Aside from the productivity impacts of specific internet applications, the value of speed to users should also be queried. This is particularly important given the goals set forth by the National Broadband Plan and related initiatives such as "100 Squared". The goal of 100 Squared is to reach a minimum of 100 million households with internet speeds that correspond to minimum download speeds of 100 megabits per second (Mbps) and minimum upload speeds of at least 50 Mbps (FCC, 2010). These speeds are somewhat controversial, because some scholars argue that the demand for higher speeds is not there and that a policy focus on speed will result in overbuilds of broadband networks (Bennett, Steward, & Atkinson, 2013). Consumer and business surveys that provide greater insight on the need for speed would help target required interventions and upgrades for reaching speed goals in a more efficient manner.

Crafting meaningful broadband policy

Many of the outstanding policy challenges associated with broadband may sound somewhat familiar to readers. Although the issues associated with speed were discussed earlier, remaining challenges include the licensing of the wireless spectrum, municipal involvement in broadband provision, competition and digital literacy. Many of these issues are ongoing challenges in telecommunications markets, dating back to the turn of the twentieth century.

The policy challenges associated with licensing the wireless spectrum date back to the dawn of the radio age and the Radio Act of 1912 (Willihganz, 1994). As detailed above, the emerging internet of things will require that additional spectrum be allocated to accommodate the burgeoning demand for resources in the broadband ecosystem. Smartphones use 24 times more data than traditional cellular telephones (Reardon, 2012). Tablets use 122 times more data (ibid.). It is yet to be determined exactly how much data a combination of smart thermostats, clothing, appliances and cars will use, but we are confident that a failure to deal with spectrum issues poses a host of negative consequences for consumers. This includes higher prices and poorer service (FCC, 2010). To meet the increased demand for spectrum, the FCC has set a goal of adding 300 MHz of spectrum between 225 MHz and 3.7 GHz by the end of 2015 (FCC, 2010). Of course, there are caveats with this plan. One of the major challenges associated with using higher frequencies in the spectrum is their limited transmission range (Wildstrom, 2012). Military radar also occupies the 3.5 GHz band and there are no plans to change this. Thus, creative solutions will be required to avoid transmission interference between military and consumer uses (Wildstrom, 2012). More importantly, innovation will be required to unlock higher frequencies in the spectrum for data transmission across larger distances. Of course, there may also be some creative ways to reallocate and re-use existing spectrum (FCC, 2010).

Aside from wireless spectrum issues, the ridiculous restrictions on municipal broadband efforts discussed in Chapter 2 remain an ongoing issue for many locations across the United States. Interestingly, in 2014, cities in North Carolina and Tennessee petitioned the FCC to overturn the laws limiting municipal broadband projects (Gross, 2015). These petitions represent growing opposition to limits on municipal initiatives; initiatives which represent meaningful pathways for overcoming the broadband divide in suburban, exurban and rural areas. Public sentiment clearly favors such efforts, as do many politicians. For example, in a January 13, 2015 speech in Cedar Falls, Iowa, which is renowned for its gigabit internet connections provided by Cedar Falls Utilities, President Obama indicated his support for efforts to remove bans on community broadband networks (Boliek & Byers, 2015). On January 22, 2015, Senators Booker, McCaskill and Markey introduced the Community Broadband Act (CBA, 2015). This act would allow local governments to build or expand broadband networks (Heaton, 2015). The recent show of support for local initiatives represents a significant pushback to big cable companies, who have been aggressively lobbying for bans on city initiatives for years (Boliek & Byers, 2015).

In many ways, the CBA (2015) and other local efforts that attempt to creatively fulfill local broadband demand reflect the growing concern about the lack of competition in telecommunications markets in the United States. In fact, the FCC has acknowledged a lack of competition in broadband markets, which is particularly evident for faster internet speeds (Brodkin, 2015). This is ironic because many of the rules and regulations supported by the FCC, at least in the past, were anticompetitive. Consider, for example, the FCC's 2005 repeal of provisions in the *96 Act*, which required providers to share their wired infrastructure with new market entrants (FCC, 2005; Brodkin, 2015). Repealing these provisions meant that new entrants into the broadband market would have to build their own networks. In many cases, this would be cost prohibitive for smaller carriers (Brodkin, 2015). This repeal, in combination with the Supreme Court's reversal of the ninth circuit U.S Court of Appeals decision that cable broadband is an information service (Tech Law, 2005) dealt significant blows to broadband competition.

Of note, however, is that on February 26, 2015, with a 3 to 2 vote, the FCC approved net neutrality rules and classified broadband service as a utility. At the same meeting, the FCC approved an order to pre-empt state laws that limit the build-out of municipal broadband internet services (Ruiz & Lohr, 2015). Not surprisingly, this order focuses on two states (NC and TN), but it is structured to create a policy framework for 21 states that have laws that restrict community broadband services. Only time will tell how this will impact competition and if private infrastructure providers find ways to fight these changes. Regardless, where competition is concerned, these rules are certainly a step in the right direction.

That said, it is important to reiterate that the anticompetitive decisions made by the FCC and many states in the mid-2000s have had significant and long-lasting impacts. Crawford (2013) lays out a compelling argument regarding the lack of telecommunications competition and its threat to national competitiveness. In short, Crawford argues that the U.S. telecommunications industry is effectively a monopoly. This means that consumers pay more for lower speed broadband access in the U.S. than consumers in other countries around the world (Gustin, 2013).

For example, in a recent report released by the New America Foundation (2014) researchers found that U.S. consumers pay more for broadband service in the 25 to 50 Mbps range than consumers in Asia and Europe. As detailed previously, there is a hope that these problems will be mitigated with the recent net neutrality decision and efforts to pre-empt state laws that limit municipal broadband efforts. Again, it will take years to before any definitive outcomes are realized.

A fourth critical policy arena that needs to be pursued to foster national competitiveness is digital literacy. Efforts to improve broadband availability will be ineffective if people do not have the ability to leverage the Internet for a variety of uses. The National Broadband Plan (FCC, 2010) readily acknowledges that digital literacy is an issue for less educated, lower income, disabled, older and minority populations. In 2010, broadband adoption rates for less educated and older Americans (> 65 years old) were only 20% and 35%, respectively, compared to the national adoption rate of 65% (FCC, 2010). These statistics highlight a disturbing trend and it suggests that some type of intervention is required. As more banking, medical and government services move online, these gaps in digital literacy become even more distressing for those left behind. Efforts to improve digital literacy will require reductions in cost of access, educational efforts to emphasize the importance and/or relevance of access for disenfranchised groups and an ability to educate those interested in participation about computers and the Internet (FCC, 2010). There are certainly challenges to overcome in this domain (Morris, 2007; Choi & DiNitto, 2013), but the costs of complacency will be more expensive and socially toxic in the long run.

Broadband in the developing world

The policy issues discussed above are not limited to the United States. In fact, the bulk of broadband challenges moving forward will stem from countries and people in the developing world. According to the Broadband Commission for Digital Development, 2.9 billion people were online at the close of 2014 and this number is expected to reach 7.6 billion in 2020 (Hill, 2014). Current statistics also highlight the tremendous potential for growth in internet adoption; people from developing countries make up 90% of people who are not yet online (UN, 2014). Mobile broadband subscriptions will be the fastest growing platform for broadband use, particularly in Africa (UN, 2014). Figure 8.1 presents the number of mobile broadband subscriptions, globally, subdivided by subscriptions in the developed and developing world. Not surprisingly, the developing world is the primary driver in mobile subscriptions worldwide.

A recent report about the digital divide in the developing world highlights a series of issues that need to be addressed to increase internet adoption rates (West, 2015). While some of these issues may sound familiar to those living in the developed world, including the cost of devices, telecommunications fees and poor infrastructure, these issues are exacerbated by extreme poverty; 25% of the world lives on less than $1.25 per day (West, 2015). Issues that are unique to the developing world include connectivity taxes that increase the cost of mobile

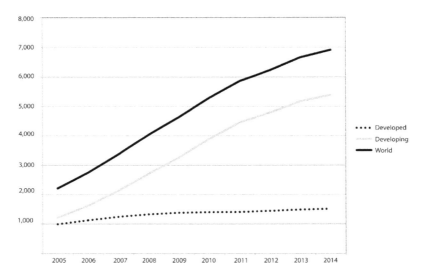

Figure 8.1 Mobile-cellular subscriptions 2005–2014 (in millions).

Source: ITU, *2014*

access, as well as censorship and a lack of content in languages other than English (West, 2015). In a different study on internet access by McKinsey & Company (2014), countries are divided into five groups based on four dimensions of inter-net barriers: 1) incentives, 2) low incomes/affordability, 3) user capability and 4) infrastructure. Across these four dimensions, countries like Bangladesh, Ethiopia, Nigeria, Pakistan and Tanzania have the highest adoption barri-ers: Internet penetration in these countries was just 15% in 2013 (McKinsey & Company, 2014). Of course, political unrest in several of these countries certainly contributes to the low penetration rates.

Clearly, there are significant barriers to obtaining internet access in the devel-oping world, many of which are distinct from the developed world. Moving for-ward, however, digital inclusion is critical. Although 40% of economic activity in developing countries revolves around agriculture, online access can provide valuable information to the digitally literate, including information about market prices, weather and disease control (Deloitte and Touche, 2014). The potential economic value of bringing developing countries online would add $2.2 trillion to global gross domestic product (GDP) and create 140 million jobs (ibid.).

Conclusion

Technology is an increasing critical facet of economic development, particularly in the Internet age, but only for those who can leverage the competitive advantage it endows on capable users. The Internet is just one of several general-purpose technologies (GPT) invented over the centuries, but it is perhaps one of the most transformative because for the first time, it allows people to communicate, in real-time, globally. This compression of space and time has unlocked a series of related innovations and new business models, the economic value of which is in the trillions of dollars. One enduring challenge in evaluating the impact of broadband on the development trajectories of regions is the ability to extract information regarding where, when, how and why broadband has enhanced regional competitiveness. Here, a spatial perspective is absolutely vital to understanding who benefits most (and least) from this GPT. While the chapters in this book demonstrate that basic availability challenges persist in the U.S. and countries around the globe, the mere availability of infrastructure alone will not be enough to enhance competitiveness. The impact that the Internet and other technologies have on individual and regional competitiveness is tied to the capacity to absorb and exploit the device and application innovations related to the Internet. Therefore, the research and policy focus moving forward must address the human component of access in a spatial context if we are to unlock the transformative powers of broadband for people and regions across the world.

Note

1 This study defines office workers as white-collar occupations that include "professionals, executives, managers, business owners, and clerical workers" (Purcell & Rainie, 2014, 6). Non-office workers are defined as blue-collar occupations such as "service workers, skilled trades, and semi-skilled workers" (Purcell & Rainie, 2014, 6).

References

Ashton, K. (2009). That 'internet of things' thing. *RFiD Journal, 22*(7), 97–114.

Baldwin, J.R., Diverty, B., & Sabourin, D. (1995). Technology use and industrial transformation: Empirical perspective. *Working Paper No. 75*, Microeconomics AnalysisDivision, Statistics Canada, Ottawa, Canada.

Benhabib, J., & Spiegel, M.M. (1994). The role of human capital in economic development: Evidence from aggregate cross-country data. *The Journal of Monetary Economics, 34*,143–73.

Bennett, R., Stewart, L.A., & Atkinson, R.D. (2013). *The whole picture: Where America's broadband networks really stand*. Retrieved from http://tinyurl.com/byn8v98

Boliek, B., & Byers, A. (2015). Obama aims to unleash broadband boost. *Politico*. January 13, 2015. Retrieved from http://tinyurl.com/mdulzff

Brodkin, J. (2015). Open Internet, but a lack of competition among providers. *New York Times*. February 4, 2015. Retrieved from http://tinyurl.com/lkbzcqa

CBA. (2015). *U.S. Senator Cory Booker introduces Community Broadband Act*. January 22, 2015. Retrieved from http://tinyurl.com/ljqvzfk

Cheng, J. (2009). Study: Surfing the Internet at work boosts productivity. *Ars Technica*. April 2, 2009. Retrieved from http://tinyurl.com/ptldorv

Choi, N.G., & DiNitto, D.M. (2013). The digital divide among low-income homebound older adults: Internet use patterns, eHealth literacy, and attitudes toward computer/ Internet use. *Journal of Medical Internet Research, 15*(5).

Crawford, S.P. (2013). *Captive audience: The telecom industry and monopoly power in the new gilded age*. New Haven and London: Yale University Press.

Deloitte and Touche. (2014). *Value of connectivity: Economic and social benefits of expanding internet access*. Retrieved from http://tinyurl.com/qj7u6l9

Entorf, H., & Kramarz, F. (1998). The impact of new technologies on wages: Lessons from matching panels on employees and on their firms. *Economic Innovation and New Technology*, 5,165–197.

FCC [Federal Communications Commission]. (2005). *FCC eliminates mandated sharing requirement on incumbents' wireline broadband Internet access services*. Retrieved from https://apps.fcc.gov/edocs_public/attachmatch/DOC-260433A1.pdf

FCC [Federal Communications Commission]. (2010). *Connecting America: The National Broadband Plan*. Retrieved from http://download.broadband.gov/plan/national-broad-band-plan.pdf

Flamm, K., Friedlander, A., Horrigan, J., & Lehr, W. (2007). Measuring broadband: Improving communications policymaking through better data collection. Pew Internet & American Life Project. Retrieved from http://tinyurl.com/qhzc634

Green Jr, P.E. (2002). Paving the last mile with glass. *IEEE Spectrum*, *39*(12), 13–14.

Gretton, P., Gali, J., & Parham, D. (2004). The effects of ICTs and complementary innovations on Australian productivity growth. In OECD (Ed.) *The economic impact of ICT: Measurement, evidence and implications*. Paris: OECD.

Gross, G. (2015). States threaten lawsuit against Obama's municipal broadband plan. *PC World*. January 26, 2015. Retrieved from http://tinyurl.com/melkmg3

Grubesic, T.H. (2008). Zip codes and spatial analysis: Problems and prospects. *Socio-Economic Planning Sciences*, *42*(2), 129–149.

Grubesic, T.H. (2015). The broadband provision tensor. *Growth and Change*, *46*(1), 58–80.

Grubesic, T.H., & Matisziw, T.C. (2006). On the use of ZIP codes and ZIP code tabulation areas (ZCTAs) for the spatial analysis of epidemiological data. *International Journal of Health Geographics*, *5*(1), 58.

Grubesic, T.H., & Murray, A.T. (2002). Constructing the divide: Spatial disparities in broadband access. *Papers in Regional Science*, *81*(2), 197–221.

Grubesic, T.H., & Murray, A.T. (2004). Waiting for broadband: Local competition and the spatial distribution of advanced telecommunication services in the United States. *Growth and Change*, *35*(2), 139–165.

Gustin, S. (2013). Is broadband Internet access a public utility? *Time*. January 9, 2013. Retrieved from http://business.time.com/2013/01/09/is-broadband-internet-access-a-public-utility/

Hartman, K. (2015). The new geographers: How Detroiters are mapping a better future for the city. *Model D*. March 10, 2015. Retrieved from http://www.modeldmedia.com/features/newgeographers031015.aspx

Heaton, B. (2015). *What impact will the Community Broadband Act have?* Government Information. Retrieved from http://www.govtech.com/network/What-Impact-Will-the-Community-Broadband-Act-Have.html

Hill, A. (2014). *Global internet users to reach 7.6 billion within the next 5 years*. September 23, 2014. Retrieved from http://tinyurl.com/lfuzqeq

International Telecommunications Union [ITU]. (2014). *ICT Indicators database*. Retrieved from http://www.itu.int/en/ITU-D/Statistics/Pages/default.aspx

JD Power. (2013). *Customer satisfaction is high among internet customers who upgrade to premium speed offerings to boost performance*. Retrieved from http://tinyurl.com/nlpj7k3

Kolko, J. (2010). How broadband changes online and offline behaviors. *Information Economics and Policy, 22*,144–152.

Mack, E., & Faggian, A. (2013). Productivity and broadband: The human factor. *International Regional Science Review, 36*(3), 392–423.

Matisziw, T.C., Grubesic, TH., & Guo, J. (2012). Robustness elasticity in complex networks. *Plos One, 7*(7), e39788.

McKinsey& Company. (2014). *Offline and falling behind: Barriers to Internet adoption.* Retrieved from http://tinyurl.com/mxfx3zw

Morris, A. (2007). E-literacy and the grey digital divide: A review with recommendations. *Journal of Information Literacy, 1*(3), 13–28.

National Broadband Map [NBM]. (2015). *Broadband test vs. advertised.* Retrieved from http://www.broadbandmap.gov/speedtest

New America Foundation. (2014). *Cost of connectivity.* Retrieved from http://tinyurl.com/khy7owh

O'Kelly, M.E., & Grubesic, T.H. (2002). Backbone topology, access, and the commercial Internet, 1997–2000. *Environment and Planning B, 29*(4), 533–552.

Purcell, K., & Rainie, L. (2014). Technology's impact on workers. Pew Research Center. Retrieved from http://www.pewInternet.org/2014/12/30/technologys-impact-on-workers/

Reardon, M. (2012). Wireless spectrum: What it is, and why you should care. *CNET* August 13, 2012. Retrieved from http://tinyurl.com/lmn8ds3

Ruiz, R.R., & Lohr, S. (2015). F.C.C. approves net neutrality rules, classifying broadband Internet service as a utility. *New York Times.* Retrieved from http://tinyurl.com/ogxegpn

Schilke, O., & Wirtz, B.W. (2012). Consumer acceptance of service bundles: An empirical investigation in the context of broadband triple play. *Information & Management, 49*(2), 81–88.

Sieber, T. (2011). 8 spectacularly wrong predictions about computers & the Internet. Retrieved from http://tinyurl.com/462yobs

Stevens, J. (2013). *Shattering the boundaries through self-efficacy: Exploring the social media habits of South African previously disadvantaged entrepreneurs* (Doctoral dissertation). Stellenbosch: Stellenbosch University.

Tech Law (2005). Supreme Court rules in Brand X Case. *Tech Law Journal.* June 27, 2005. Retrieved from http://tinyurl.com/ptgyz4p

UN. (2014). Internet well on way to 3 billion users, UN telecom agency reports. *UN News Centre.* Retrieved from http://tinyurl.com/mrjgqua

Van der Krogt, J. (2011). *The influence of social media on nascent entrepreneurship in the Netherlands* (Master Strategic Management Thesis submitted to Tilburg University). Retrieved from https://prezi.com/mudeezeavk-k/the-influence-of-social-media-on-nascent-entrepreneurship-in-the-netherlands/

VantagePoint. (2015). *Wireless broadband is not a viable substitute for wireline broadband.* Retrieved from http://tinyurl.com/kr8g9hl

Websense (2005). *$178 billion in employee productivity lost in the U.S. annually due to Internet misuse, reports Websense, Inc.* Retrieved from http://tinyurl.com/ll46za6

West, D.M. (2015). *Digital divide: Improving Internet access in the developing world through affordable services and diverse content.* Center for Technology Innovation. Brookings Institution. Retrieved from http://tinyurl.com/n56cna7

Wildstrom, S. (2012). New approaches to spectrum: The challenge of wireless data. *Cisco.* Retrieved from http://tinyurl.com/kdtm8n3

184 The future of broadband

Willihnganz, J. (1994). Debating mass communication during the rise and fall of broadcasting. *BRIE Working Paper 74*. Berkeley. Retrieved from http://brie.berkeley.edu/publications/WP%2074.pdf

Wilson, M.I., & Corey, K.E. (2011). Approaching ubiquity: Global trends and issues in ICT access and use. *Journal of Urban Technology, 18*(1), 7–20.

Worstall, T. (2013). The beginning of the death of wired broadband. *Forbes*. Retrieved from http://tinyurl.com/k9dr6ya

Zook, M., Graham, M., Shelton, T., & Gorman, S. (2010). Volunteered geographic information and crowdsourcing disaster relief: A case study of the Haitian earthquake. *World Medical & Health Policy, 2*(2), 7–33.

Index

For Product Safety Concerns and Information please contact our EU
representative GPSR@taylorandfrancis.com
Taylor & Francis Verlag GmbH, Kaufingerstraße 24, 80331 München, Germany